SNIPING, SCOUTING AND PATROLLING

A textbook for Instructors
and
Students

PRICE FOUR SHILLINGS
(*per post* 4/3)

GALE & POLDEN LIMITED, ALDERSHOT

Published by S.O.A. Books, Canberra
First published by Gale & Polden Ltd. in 1940

ISBN: 978-1-925907-14-8

This edition © S.O.A. Books, 2024, all rights reserved. No part of this publication may be reproduced, stored in a retrieval system or transmitted, in any form or by any means, electronic, mechanical, photocopying, recording or otherwise without prior permission from the publisher.

Printed in Great Britain by
Gale & Polden Ltd.
Aldershot

P-8138

CONTENTS

CHAP.		PAGE
I.	THE SNIPER'S ART AND QUALIFICATIONS	1
II.	ORGANIZATION OF SNIPERS AND SCOUTS	10
III.	EQUIPMENT	14
IV.	DAY WORK	17
V.	NIGHT WORK	23
VI.	USE OF THE SCOUT-SNIPER UNDER VARIOUS CONDITIONS	27
VII.	SNIPERS IN ATTACK AND DEFENCE	32
VIII.	INTELLIGENCE	38
IX.	NOTES ON THE COLLECTION AND TRANSMISSION OF INTELLIGENCE BY TROOPS	42
X.	REPORT WRITING	44
XI.	THE PRISMATIC COMPASS	47
XII.	THE PROTRACTOR	57
XIII.	MAP READING	59
XIV.	OBSERVATION OF GROUND	74
XV.	SNIPING POSTS	79
XVI.	STALKING AND THE USE OF COVER	87
XVII.	THE LOCATION AND CONSTRUCTION OF SNIPERS' AND OBSERVERS' POSTS	91
XVIII.	THE CONSTRUCTION OF LOOPHOLES	103
XIX.	DECOYS AND THEIR USE	113
XX.	SQUIRT GUNS	116
XXI.	THE TELESCOPIC SIGHT	119
XXII.	CONCLUSION	123

LIST OF PLATES

PLATES		PAGE
I.	THE RIGHT WAY TO MAKE USE OF COVER	7
II.	THE CORRECT USE OF COVER AND LOOPHOLES	20
III.	THE WRONG WAY TO SHOOT THROUGH A WINDOW	29
IV.	SKETCH OF FRONT WITH RANGES	39
V.	SYMBOLS USED ON ORDNANCE MAPS	60
VI.	ENLARGING A MAP	66
VII.	DEFINITIONS EXPLAINED	68
VIII.	TRAVERSING	70
IX.	TRAVERSING	72
X.	SNIPING POSTS	80
XI.	SNIPING POSTS	81
XII.	SNIPING POSTS	82
XIII.	RANGE CARD	84
XIV.	VIEW FROM LOOPHOLE, SHOWING RANGE CARD	85
XV.	REAR VIEW OF INTERIOR OF LYING SNIPING POST	92
XVI.	REAR VIEW OF STANDING SNIPING POST	94
XVII.	SECTIONAL VIEW OF LYING POST	96
XVIII.	STANDING POST	98
XIX.	SNIPER WITH P. '14 RIFLE AND TELESCOPIC SIGHT	125
XX.	FRONT VIEW OF SAME MAN	126
XXI.	CAMOUFLAGE SCREEN	127
XXII.	CAMOUFLAGE SCREEN SEEN FROM FRONT	127
XXIII.	CAMOUFLAGE MADE OF NET AND RAGS	128
XXIV.	CAMOUFLAGE OF NET AND BRACKEN	128
XXV.	MEN IN CAMOUFLAGE NETS	129
XXVI.	SAME MEN SEEN FROM A DISTANCE	129
XXVII.	SHOWING USE OF CAMOUFLAGED HEAD	130
XXVIII.	SNIPER MAKING USE OF NATURAL COVER	131
XXIX.	TYPE OF DISAPPEARING TARGET, USEFUL FOR SNIPER PRACTICE	131

ACKNOWLEDGMENTS

JUST as this book was being completed for press two excellent articles appeared in *The Times*, and by permission of the Editor of that paper have been incorporated in the text.

The first, " The Sniper's Art," has been used as introductory matter to the first chapter. The second, " Squirt Guns," forms a separate chapter.

The publishers are indebted to the *N.R.A. Journal*, official organ of the National Rifle Association, for permission to use the photographs which appear as Plates XIX, XX, XXVIII and XXIX.

one

THE SNIPER'S ART AND QUALIFICATIONS

IN August, 1914, the Germans had in each infantry battalion a number of trained Snipers. In the armies of the Allies there was no such organization and the whole subject, together with the best methods of training, had to be learnt and organized from experience. To-day the Germans are reported to have forty Snipers in each infantry battalion. The Allies have also their Sniping organization and behind them is all the experience gained during 1914-1918. That experience is embodied in this book, together with some recent developments from men who have been working and thinking about the subject ever since.

Amongst those uninstructed in the art there still exists the fallacy that all that is necessary to produce a Sniper is to take a battalion or company shot, or if possible a King's Prize or King's Medal winner, and equip him with a rifle fitted with a telescopic sight. In fact, though expert marksmanship is a necessity in a Sniper, it represents not more than about thirty per cent. of his required mental and physical equipment.

The Sniper is first and foremost a hunter and, in open warfare—and, it may be added, for work in No Man's Land—the art of stalking has to be developed to a high degree of excellence.

The Sniper with any sort of static position from which to work may and does construct a hide just as does the bird photographer and, like him, may occupy that hide with infinite patience for many hours before getting a chance of a shot. On the other hand, like the big game hunter, he may have to go after his quarry, pitting his wits against the enemy, suspecting and watching every unusual appearance in the ground or vegetation and using his eyes, ears and nose every instant he is out.

The work of the Sniper is closely allied to that of the Scout, and in fact one man often does both jobs. In most armies he is part of the front-line intelligence organization and part of his duty is the gathering of information. There is, however, this difference. The Scout is solely concerned with information and getting it back in the quickest possible way, whereas the Sniper will usually act on his own information if he can do so to the discomfort of the enemy. If, for instance, he can discover a machine-gun nest or the whereabouts of an enemy Sniper, he will do his best then and there to put them out of action. The Scout with similar information would send it back at once for the use of others. The Scout-Sniper works as he is ordered. He may be sent out for information only, or he may be required to get his man. He suits his method to his job. In any case the gathering of information is always part of his work.

One of the chief things the Scout-Sniper has to learn is the art of concealment. His dress must enable him to match with the country through which he is moving. He dirties his face and hands, for in almost any light flesh shows up in a surprising manner. He dirties his rifle also and takes care to hide the object-glass of his telescope by means of a long cover. He learns to stay absolutely still for long periods and how to crawl an inch at a time using only his hands, or possibly one hand, to pull himself along. He learns to make hides so cunningly that he has a wide arc of view from something that looks like a natural feature of the landscape.

A good Scout is not of necessity a good Sniper, but a Sniper must be a first-class Scout. The best Snipers and also the best Scouts are usually country-bred men, but it does happen that men from the towns develop a flair for the game and become second to none at it.

To be of any use the Sniper must be an enthusiast. Detailing men to Sniper duties as one would to a

THE SNIPER'S ART AND QUALIFICATIONS

" fatigue " is fatal. Snipers must be chosen from those who wish to do the work and have proved themselves capable of it under instruction.

The Sniper should be known as a good shot before he comes under special instruction. Then all he will have to learn about this part of the game is how to use the telescopic sight and the details of his special rifle if one is issued. These and some actual shooting at special targets should complete his training as a shot.

Following are the qualifications for a Scout-Sniper :—

1. Must be sober and of temperate habits.
2. Must be physically fit for great strain and hardships.
3. Must be a good shot at fleeting targets.
4. Must know how to do the right thing at the right time, and to do it quickly.
5. Must be a keen observer ; must be able to see without being seen.
6. Must know how to stalk without being seen or heard.
7. Must know how to take advantage of natural cover, and how to construct artificial cover.
8. Must have and cultivate great patience.
9. Must be a good judge of distance.
10. Must have keen eyesight and good hearing.
11. Must know how to use telescope, periscope, and telescopic rifle.
12. Must understand the use of the compass.
13. Must be able to read a map.
14. Must have great self-control.
15. Must have strong will-power and determination.
16. Must be able to write short and concise reports.
17. Must be able to sketch.
18. Must know how to make and use decoys.
19. Must know how to locate enemy Snipers, machine guns, and artillery by flash, sound, and bullet strike.
20. Must be a man who is a sticker.

21. Must be a man with courage combined with coolness.
22. Must be a man who is not slovenly, fussy, nor one who panics.
23. Must be truthful in all circumstances.

After reading these qualifications the student may be lead to believe that nothing less than a superman will satisfy the Officer i/c Scout-Snipers. In fact, a large number of the qualifications above enumerated may be learned by any intelligent man. Intelligence really is an essential. A dullard is no good at all for the work.

An essential which must be insisted upon is truthfulness. Reports of observations and targets hit must be exact. Men will lie to get a good reputation, to save themselves trouble or, in some cases, because they have vivid imaginations and jump to conclusions which are not warranted. This last class may also be conscientious and in training will correct reports that have previously been sent in. Imagination is a very useful thing and perhaps it is true that a man without imagination will never make a Scout or a Sniper. In such a case reiterated warnings that the word " suspected " should be used unless a fact is ascertained will be sufficient to correct any tendency towards exaggeration.

Observing is a very fatiguing job, even from properly constructed posts where every possible convenience for comfort exists. Towards the end of a two hours spell the best of men may tend to be " fed up." Those responsible for training and particularly Intelligence Officers or others in charge of Scout-Sniping should be on the lookout for inaccurate reports in such conditions.

Good eyesight is of course a necessity, but unless the student is a hunter or poacher or otherwise educated in watching natural objects and in actual " seeing " the best eyesight needs considerable education. It is not so much what the eyes see, but the message they send to the brain which is of importance.

Men who must wear glasses to correct faulty vision should not be used as Snipers, Scouts or Observers, even if their corrected vision is super-excellent. If

THE SNIPER'S ART AND QUALIFICATIONS

sunlight is reflected from their glasses they are a danger to themselves and others. If they lose or break their glasses they are out of action.

Some men can see very well indeed by day but are partially blind by night. Night blindness might be thought to prevent a man from being a good Scout or Sniper, but in some circumstances a man may be so good by day that it is worth keeping him for day work only even if he cannot see at night. Colour blindness should be regarded as an absolute bar in Scouting or Sniper work. Colour, either on uniforms or in natural objects, is often of the utmost importance in making observations.

First-class hearing over the whole range of audible sounds is another absolute essential. At night a Scout's ears may often be of far more importance than his eyes. Again education is necessary. The Scout-Sniper must know the meaning of every sound his ears register.

The ability to sketch is one of the qualifications set forth in the list we are discussing. Note here that nothing like an artistic production is required. A field sketch is a very simple thing which shows only the essentials, and conventionalized forms can be used to indicate everything that is in front of the observer. Most men can learn to make useful field sketches after a few hours of instruction and practice. This subject will be further dealt with.

The first qualification set forth is sobriety and temperance. Sobriety refers to a man's general habits. He should not be of an excitable temperament. Temperance is used with its ordinary accepted meaning, indicating a close watch on all habits tending to lower both mental and physical activity. Alcohol and nicotine are narcotics acting on different nerve centres, but tending to dull the senses if used in any but the most moderate way. A heavy drinker is of course impossible as a Scout-Sniper, for he will always be unreliable and usually incapable of the heavy mental and physical strain necessary to his job. The heavy smoker is not likely to be mentally dull but he will be physically of a lower standard than the non-smoker or very moderate smoker. Smoking and drinking when at work are absolutely forbidden to the Scout-Sniper and one of the dangers in employing a heavy smoker is that he becomes depressed when deprived of his customary solace and is more quickly tired during a long spell of work.

The non-smoker can often cultivate a very keen sense of smell which will be useful to him in observation both by day and by night.

Sniping is extremely interesting work. The keen man delights in matching himself against a cunning and experienced enemy, in being top dog, and staying that way. He takes all the advantages he can and gives none. If by any chance the enemy Sniper sees him he gets him first. A good Sniper-Scout will always be thinking about his work, dream about it at night, invent all kinds of decoys and contrivances to conform with the local conditions.

The Scout-Sniper is always learning. No one knows everything, though some may think they do. There are hundreds of big and little things about this work yet to be found out, invented, and to be put into actual use before this second great war is over. Keep your mind busy on your work, and in connection with your daily practical experiences; you will be benefited to-morrow by your experiences of to-day. Experience may give negative as well as positive results. Always bear in mind that when you are trying to locate the enemy Sniper, he is at the same time trying to locate you. Snipers belonging to the same unit should constantly discuss ways and means amongst themselves—two heads are better than one—but they should never discuss these matters with strangers. If you have some ideas that will benefit other battalions, it will be communicated to them through the proper channels.

The Sniper's duty is to worry the enemy night and day, give him no peace at any time. He is always on the offensive. The least part of a man showing himself by day should be fired at, unless by waiting a little he

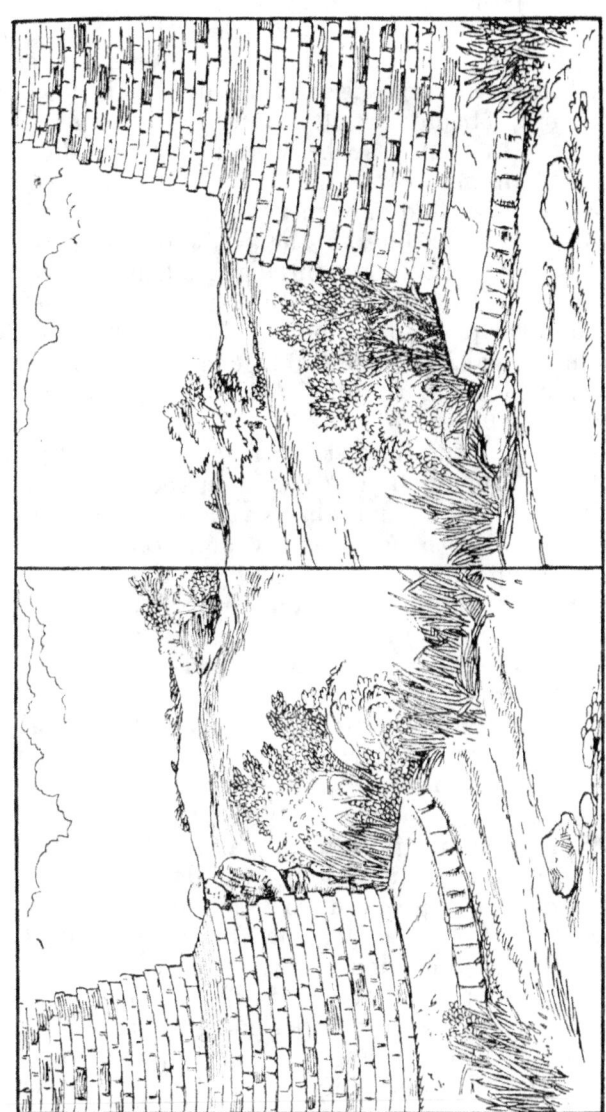

When out Scouting always lie down to look around a corner, taking advantage of all Natural Cover, and whenever possible look from your right or Firing side

will expose himself more, giving a better chance to put him permanently out of action. This does not mean uselessly blazing away, which, of course, tends to give away your post or position.

The ideal of the Sniper is to kill with one shot any enemy he sees. He may, however, in some circumstances, have orders to keep the enemy heads down, as for instance, if Scouts are after a sniper post or machine-gun nest. In that case he will fire at the slightest movement or may even fire at the indicated post or nest when he sees no movement at all.

Scout-Snipers belong to the battalion and usually work under the Battalion Intelligence Officer. In the present-day establishment each company provides an equal number of Snipers and should be responsible for replacing casualties. The Intelligence Officer is usually responsible for the training in the field of men who replace casualties.

Nothing but the highest efficiency in shooting and trench-craft should be considered in the appointment of a Sniper. Sniping and observation posts can always be improved, both as to safety and comfort. Take pride in leaving something behind you that will puzzle your successor to improve upon. But do not be hasty. Give your predecessors some credit for thoughtfulness and common sense. Nothing is likely to annoy you more in returning to forward defence localities than to find that your most cherished post has been given away by over-zeal on the part of those you are relieving. Remember this and be quite sure of your reasons before making any changes. In making improvements, on advanced Sniping Posts in particular, great care must be taken that, in so doing, you don't overdo anything and thereby give your position away.

For all that it is a fact that the old saying, that " What was good enough for my father is good enough for me," is a curse in modern warfare. We live in an age of progress. We are either going to be best or the enemy is. Take advantage of everything at your disposal, with an eye to improvement. Explain thoroughly to the Sniper who relieves you, after showing him all the ins and outs of your improvements, your reasons for making them. We have no trade secrets or patents, except as far as the enemy is concerned. Any idea you may have let your brother Sniper benefit by it. Nothing but united effort will lead to success. Team work, and team work all the time, is the only way to win. Many lives were needlessly thrown away in 1914-1918 by jealousy leading to lack of co-operation.

two

ORGANIZATION OF SNIPERS AND SCOUTS

[NOTE.—*This chapter represents the ideal of the writer, who had great experience in Sniping and Scouting and in training Scout-Snipers during the last war. It contains much sound common sense, but in reading it the student should remember that it represents an ideal and may not conform in every particular to present practice. No official information on the present Sniper organization is available for publication, but it is known that there is a proper organization in preparation which in some cases is actually at work and is being extended throughout the Army. It is believed that all Scouting and Sniping work comes under the Battalion Intelligence Officer, whose duty it is to forward reports daily to Brigade and that Brigade deals with these reports in a manner laid down by regulations so that eventually all useful matter reaches Corps. Scouting, Sniping and Observation organization should always be fluid so that it can be adapted to the needs of the moment. Any rules laid down or any suggestions made in this section should be read with these reservations in mind.*—THE EDITOR.]

A SPECIALLY appointed officer in the Army Corps Staff should have control of organization, employment, and disposition of all Snipers in the Corps. There should be also one Divisional, one Brigade, and one Battalion Sniping Officer. This is being successfully done now by the Intelligence Officer.

Snipers must form a separate unit in the battalion, and should live apart from their companies, should **never** be employed in any ordinary work, and should wear a special badge. The officer in charge is responsible that they are kept to the standard with regard to discipline. A Sniper is a specialist, but only when he is so employed. He must not forget that he must at all times be amenable to military discipline.

ORGANIZATION OF SNIPERS AND SCOUTS

The strength of Snipers in the battalion should be as follows :—

    1 Officer.
    8 N.C.Os.
    32 Men.

No preference as to who should be a Sniper must in any case be given ; nothing but qualifications must be considered. Any man selected for a Sniper must at least be a first-class shot. The same qualifications must apply to the Officer and N.C.Os., as it would be a crime to put such highly trained men under anyone who is superior to them in rank only. The best man is the volunteer.

If Snipers are employed before regular Sniping organization has been adopted in the battalion, they should be left to their own resources, given liberty of action, live with their company, report daily to the Company Officer through the proper channel, but should be exempt from all other duties. They should work in pairs or threes. In regard to discipline, pay and rations they will form part of their company. None but fully qualified Snipers should be employed in this way.

There should be two posts on each company frontage, with three men manning each, and two spare men will have a roving commission along their company frontage, under the O.C. Scout-Snipers' directions.

When posts are accessible in the daytime, reliefs should take place every two hours. This is done by the Rovers or the support company Snipers, and only one man should be relieved at one time from each post. That will relieve each individual every four hours. O.C. Snipers must see that they are always on the alert. This is an all-the-time job ; every minute counts. A Sniper falling asleep or otherwise neglecting his duties should be severely punished.

A detached advanced post where no communication can be maintained in the daytime should have a phone or signal wire, so O.C. Snipers can readily obtain information. The senior N.C.O. or Sniper in command of post will be the one responsible for writing the daily report, and will be held responsible for general conduct and efficiency of that particular post.

Any casualties must be included in the daily report, with full explanations and details. Any Sniper putting in a wrong or exaggerated report should be severely dealt with.

It may, under certain special conditions, be necessary to have more or less than six fixed Sniping Posts in connection with the three companies occupying the front line of trenches. In that case some or all of the support company Snipers may be used. It would be well to have several spare Sniping Posts under ordinary conditions. Snipers can then move about, and not be shooting continuously from the same place. On special occasions the spare posts would be useful for the support Snipers. In other cases good work might be done from the support trenches. The two men who are Rovers to-day will each relieve a man in the fixed Sniping Post to-morrow. That means that each Sniping Post is always manned by one man that was there the day before. This being the case, the battalion that relieves you must send in one man for each Sniping Post the day before the battalion takes over the trenches. These advanced Snipers report to, and come under the direction of, the present O.C. Snipers, but make their daily reports to their own O.C. Snipers when he takes over the next night. O.C. Snipers must make himself perfectly familiar with each Sniping Post under his command. He must also know the individual ability of men under his command, and employ them accordingly. Some men shoot quicker than others; such men are best employed on the shorter range, where the enemy show themselves for a very short period. PUT THE RIGHT MAN IN THE RIGHT PLACE.

A diary must be kept of every Sniping Post by O.C. Snipers, and should consist of all the Snipers' daily reports. The O.C. Snipers must give a receipt in writing when taking over relieved battalion's diary, and *vice versa*. All O.C. Snipers must keep a copy of

ORGANIZATION OF SNIPERS AND SCOUTS

diary, and have one sent to Battalion and Brigade Headquarters. This diary must be at all times accessible to the Intelligence Department, also to the artillery observer; but no one must at any time read the report, or any part thereof, without properly establishing his identity. Anyone allowing a stranger to be in possession of or read any part of a Sniper's diary should be severely dealt with.

If these rules are strictly observed it will be of great value in many ways. Scout-Snipers should also receive any information procured by airmen, artillery observer, etc., that will be of any value to them. CO-OPERATION IS VERY ESSENTIAL.

three

EQUIPMENT

HOWEVER efficient a Scout-Sniper may be, if he has nothing to work with, or insufficient equipment, both as to quality and to quantity, he is severely handicapped. This branch of the Service is of sufficient importance to warrant the best of everything. We demand the highest possible qualifications in the man selected for this work, and it is but fair to him that we should give him the best equipment obtainable.

Some years ago each infantry battalion had issued to it a number of Pattern '14 rifles fitted with telescopic sights. It is believed that these rifles and sights are now the Sniper's standard equipment. The Pattern '14 rifle is an accurate and sturdy arm and well suited to Sniping. It was originally designed to take a .276 inch calibre cartridge and was to have been the new Service rifle. The Great War broke out before it had been finally passed for manufacture. During that war there was a demand for large numbers of rifles and the United States took over the supply contract. The action of the Pattern '14, being of the Mauser type, was far easier to make than our rifle, Short Magazine Lee-Enfield, and so the .276 was redesigned—a matter of very little trouble—to take the Mark VII .303 inch cartridge. Very large numbers were manufactured in the States.

The woodwork, having been quickly assembled, is not always properly seasoned and a watch should be kept for warping fore-ends and handguards. The barrels are sometimes on the soft side and are therefore liable to wear quickly. A watch should always be kept for this if in fact the Pattern '14 is the Sniper's arm.

The telescopic sights are of the Aldis pattern and are remarkably good for the work. These sights will be discussed under a separate heading.

The telescopic sight is not altogether suited for night shooting, at any rate in very dark conditions, and some provision should be made for fitting to the rifle large and easily visible sights for night work. This also is discussed under a separate heading.

The Sniper's rifle must be made as invisible as possible by painting or binding the barrel with black sticky tape of the kind used by electricians. If the rifle has a bright yellow butt or very light-coloured fore-end, paint should always be used. The barrel may be painted grey. The bright muzzle of a new rifle might give your position away if not painted. Under many circumstances green would be a better colour. It will also in many cases be of advantage to have the whole rifle painted, telescopic sight and all.

Each Sniper must have his own rifle, for no two men find their zeros exactly the same. A man takes a pride in his own rifle and will take the greatest possible care of it. It should be his responsibility and his alone to see that it is always in perfect order.

The O.C. Snipers must understand how to adjust telescopic sights. Stripping down, or anything except the adjustment of focus, sighting and lateral direction, is forbidden. The Sniper will make his own elevation changes and a good man can be trusted to make the other two allowed adjustments as well.

A telescope or a pair of field-glasses is needed. A powerful telescope is better for this work than field-glasses, though the wide field of view given by a good binocular is occasionally very useful when it can be used without danger of exposure.

The issue of compasses to Snipers will depend on the work they have to do. A compass should be regarded as essential in any observation post as bearings are important in locating points under observation. The use of the compass is discussed in a later section.

A periscope may or may not be useful to the Scout-Sniper. The situation and construction of the post or hide will govern this. The equipment of the section should include sufficient periscopes for all possible needs. Periscopes are always useful to men working from trenches.

An abundance of camouflage material is needed. Hoods, smocks and crawling suits in various shades of green, or other colours to conform to the surroundings, should be supplied by the Brigade. These coats will be found very useful to assimilate to the surroundings, and it is usually well to ask for them plain, with a supply of paint to colour them to suit local conditions. Badly coloured clothing is dangerous. An uncoloured coat or smock is better than a badly coloured one. Veils and masks to suit, and covering for the hands of the same material should be made. It is wonderful what stalking an expert can do when assimilated to the surroundings. The properly camouflaged Sniper or Scout can occupy an advanced and exposed post without fear of detection. A grass head-cover is very useful when well made.

Other material to be indented for by O.C. Snipers is as follows :—Ropes for climbing trees (with big trunks) for placing Snipers or decoys. Wires of different gauges to manipulate decoys, and loophole covers. Paintbrushes and ready-mixed paint of colours to suit requirements for rifles, coats, etc. Steel plates for making loopholes. Sandbags for constructing special Sniping Posts. A few yards of factory cotton for making faces of decoys. Needles and strong thread. Wire netting, lumber and timber for constructing posts. Old clothing, caps, and boots for dressing decoys.

Wire cutters, revolvers and signal pistols and lights are also part of the section's equipment. Boxes for storing equipment are also needed.

four

DAY WORK

THERE is much for a Scout-Sniper to do by day, as you will see later. The kind of work he does varies. With Snipers the total number of the enemy killed is not the great thing, though very important. It is upon the Scout-Sniper that the Brigade relies for information and reports. The information is asked for, and it is up to the man to find the means by which it can be obtained.

There are various posts and plans from which information can be found. Posts are divided into three kinds :—

    (1) Organized.
    (2) Unorganized, but fixed.
    (3) Moving or Rovers' posts.

An Organized Post is a post dug in, whether in No Man's Land, in the parapet, in the parados, in the reserves, or anywhere which is permanent. Such a post is used by Snipers daily.

The Unorganized Fixed Post might be well dug in, but used at various times, not continually. Of course, this keeps the enemy guessing as to the Sniper's whereabouts. If you spotted an enemy's unorganized post in No Man's Land, what would you do ? The enemy will do the same.

The moving Sniper, known as the Rover, wanders around the trenches, hedges, shell-holes, and old buildings, etc., and takes the pot shot. There has been great argument as to using buildings as posts. There were buildings on the front in the Great War which we could use with comparative safety, and there were others which it would be suicide to go near. This is where the common sense of the individual Sniper comes in. If you do use a building for a Fixed Post, do not hesitate to sandbag it well in. The construction of Sniping Posts will be described later.

Having manned a post, we say to ourselves, What instruments are necessary? The answer to that is: Telescopic sight and rifle, periscope, field-glasses and telescope, and perhaps a dagger-knife, a knife of this kind being a silent and useful weapon for night work.

Fixed Posts should be manned by two or three men, as constant observation is very trying. Allow one man to observe for twenty minutes, one man to shoot, and one man to act as runner, the runner being responsible for messages back to Headquarters. If the post is in No Man's Land and your report cannot be taken back during daylight a runner is unnecessary. It is possible to devise methods of signalling, but lamps, telegraph keys and telephone should be used with great discretion. They all make a noise and messages sent by their aid may be read by sound. Noise is one of the great enemies of the Scout-Sniper.

From the observation hole of the Fixed Post a detailed range card and sketch should be made, upon which must be described all minute details on the sector, such as shell-holes, upturned sods, craters, old tins, sandbags, etc.; anything noticed must be constantly watched and continually observed, and reported back.

The sector or frontage of the post should not cover a wide arc. The Observer will not then have too difficult a task. By constant watching and continuous study of No Man's Land, any change will be readily noted. Both Observer and Sniper should know their sector "like the backs of their hands."

Remember, persistently hammering at one spot tells tremendously, but remember also that watching is much more important than uselessly blazing away.

DAY WORK

Persistent hammering may mean firing a shot at a known target once every half-hour.

After concealing himself in the post or in the roving position, the Sniper must conceal his fire from the enemy. This is quite easy to understand, for the flash of a single shot might give away a post which took weeks to construct. If the country is covered with gorse, a bush can be put in front of the loophole and shot through, thus hiding the flash. The cover in front of the loophole would correspond with the surrounding natural features. The rifle should be well inside the hole if possible. Two feet is an ideal to be worked for. The Fixed Post should have front, rear and overhead cover, and means should be provided for firing at an angle; not only will loophole be less likely spotted, but flash less likely seen. Have the background inside the post correspond with that in the front. It is far easier to see through a loophole with a white background than one with a dark one. In all cases beware of the silhouette effect.

Each and every man must have patience. If it is necessary to take ten minutes to fire a shot, take it, and don't forget to always have a reason for shooting, such as unnatural movement of bush, etc. If fire is opened at anything which looks suspicious, two or three shots at fairly close time intervals may be effective. If an enemy Sniper is actually there, you may kill him, but if you don't hit him, you may make him think that you have definitely located him, and he won't fire from there any more. In this connection, always bear in mind that all unnecessary shooting should be avoided, as it may give your position away. BE SUSPICIOUS; ONCE SUSPICIOUS, YOU WATCH; ONCE YOU WATCH, YOU LEARN.

As to the kind of information wanted about the enemy the following are suggestions. Loopholes in any part of his trenches or otherwise; unnatural movements; Sniping Posts in trees, hedges, farmhouses, craters, shell-holes, etc.; information about his stores, trenches, wires, activity, dumping ground, gaps in

PLATE II.

NEVER look over Rails as above, but dig down and shoot between the Ties or Sleepers. Tunnel through Rail Road from opposite side

Always keep Muzzle of Rifle well back from loophole
NEVER Shoot as shown

DAY WORK

parapet; points from where he can best be sniped; fires where cooking goes on; fixed rifles, machine guns, bombing points, saps, uniforms, habits, etc.; reports on aeroplanes or aircraft of any kind; direction from and where going (remember that time is imperative); enemy's guns (calibre), and, roughly, direction. Wind is most important in these barbarous times, and the slightest change must be noted and reported at once, if possible. These are some of the things to be noted, but there are more; something new arises daily, so watch hard, and, as we have said before, patiently. The duty of getting special information is usually given to a small patrol of which Scouts are the whole or a part.

Information about our own lines is important. You may ask yourself why. If you have your post in No Man's Land, can't you see our lines as the enemy sees them? Then report fully in what places improvements can be made. Suggest new points for observation and Sniping. Report the dangerous places in our line where the enemy snipe, and give the direction. If it is out of your sector, pass the information on to the post responsible; if not, it is up to you to deal with the enemy Sniper. Report anything that will be of use in the perfecting of your own front; it is marvellous the good such information as this does.

Man your post with the same men until they thoroughly know their frontage; each inch should be known. It is important that Scout-Snipers should know thoroughly their whole battalion frontage.

It is important that the training of eyesight, not only the naked eye, but with the telescope, should be well advanced before the men go into the front line. Students are advised to put in some time each day in observation. The more you look at one spot the more you see. All movement is suspicious—a branch apparently blown by the wind, the flight of birds, etc.— so watch every movement hard. Your sector should be under constant observation every minute of daylight. At first you may think you have seen everything; don't fool yourself, there is something new to see every day, perhaps every hour or minute.

A Fixed Post should be manned before dawn, and not quitted until after dark. It is unnecessary to tell you that once your post is discovered by the enemy, it is of little use to you, so don't fire too often, or it will disclose your position. This is where patience is taxed. Don't be in a hurry to fire; make sure of your target first, figure on 50—50, and if you are a good shot you will seldom lose. A good plan is not to fire for some time; that will entice the enemy to come out; he will get careless, then the odds are yours.

Build dummy loopholes, not too many, in your trenches; have them marked, and let the enemy play with them. It will keep him guessing whether they are manned or not, and, as will be explained later, will give you a chance to locate enemy positions. Give them a few shots from dummy positions now and then to encourage them to waste ammunition. Always remember that, though you are trying to get the enemy, he is at the same time trying to get you.

In ordinary every-day trench warfare it is a good rule that nobody but the Snipers should fire. Of course, if there was an attack or daylight raid, matters would alter, but otherwise it would be a saving of ammunition with about the same results. All advanced Sniping Posts should be manned and guarded at night, and can nearly always be used with advantage as listening posts. It is dangerous to leave them unoccupied, as the enemy might find them and mine them, or have the other branches, such as machine guns, trench mortars, artillery, etc., surprise you in the morning.

Co-operate with the F.O.O. (artillery); learn what he knows; also with the Air Service, and obtain the use of photographs from aeroplanes. These will aid your findings wonderfully well. Any information received should be immediately signalled to you if that is possible. Certainly the Scout-Sniper should have all possible information at the earliest moment.

five

NIGHT WORK

LET us divide night work into two parts—(1) Night Sniping and Observation from Posts ; (2) Patrol Work.

If you believed that you could snipe the enemy by night by the help of star-shells, either their defects in parapet or working parties, then, of course, you would do so ; but remember, if you have a good post in No Man's Land, do not foolishly give it away by attempting night work, for the flash of a rifle is much easier seen by night than it is by day. If you are so well concealed that the flash cannot be seen, then, of course, it would be all right.

Night is the time for the Roving Sniper; he can crawl out to shell-holes, depressions, mine-craters, etc., and spend a jolly evening playing with the enemy working parties.

Before going out on patrol for information there are many things to think about. Nothing is so foolish as to go aimlessly over the parapet.

No patrol will ever go out without orders or without a definite task. The task should be limited. If more information is required than can be got by one patrol another should be sent out, but not at the same time. A patrol out for information must be prepared to fight for it, but is really successful if it gets it without fighting. A dog-fight puts the enemy on his guard, which may be a thing at all costs to be avoided.

Suppose you have been told to take out a patrol to obtain certain information from a certain sector about the enemy's wire, its condition, depth, etc., whether it

SNIPING, SCOUTING AND PATROLLING

has been cut or not. Your patrol would consist of about three or four, no more; it is information you want, and this number is plenty.

It is dangerous to go out without warning your sentries from what point you are going out, and time that you are returning; warn all sentries yourself. Do not be too definite about the time you will return. Tell them to expect you at any time after a certain hour. You may find unexpected difficulties and be delayed.

Think of the ground that you are going to cover; you have been studying it by day by periscope or through loophole, and know every detail; you have pictured the places you are going over the parapet, the position where you are going through your own wire, and how you are going to get across No Man's Land; the lie of the land, the disposition of your men in crawling, etc.

In ninety-nine cases out of a hundred you will be successful; but supposing one of your men knocks a tin or rings a bell attached to a wire, the result would be continuous lights or star-shells. Then lie still and flat, face and hands covered. If caught standing and the star-shell bursts, remain perfectly rigid, and lower head and place hands slowly behind back; if you are seen then, and you know the position of a shell-hole, get to it as quickly as possible.

While on patrol duty never hurry; crawl slowly and quietly, heels well down, face and hands covered with paint or mud. The skin is what shows up. For every look in front give ten behind; of course, this is exaggerated, but it is to impress on you the extreme importance of keeping alert. When looking back, do not raise your head above your right or left shoulder, but raise your right or left upper arm on a level with the shoulder, and look back under the armpit; by this means nothing will protrude higher than the original thickness of the body. If four men go out one can watch the front—the leader—one the rear and the other two their own flanks. Much drill is necessary for

NIGHT WORK

night patrolling. The men should be close enough to keep in communication by sight or touch. If one stops all should stop on the instant. The leader will be the guide for ordinary stops and for direction.

When in No Man's Land, look back and spot all land-marks, and mark them in relation to the point where you came out of the trench; if it is too dark to do that, feel the condition of the earth at that point—it might tell you something; or put a notch in the entanglement picquet, and remember the number of yards left or right of that. If you don't do these things, you may find yourself in No Man's Land one day with dawn breaking, and it will be necessary for you to crawl into a shell-hole and spend the day. Never move more than ten yards without lying perfectly still and listening for at least two minutes; by doing this, you know if an enemy patrol is out, or you can hear enemy wiring parties, if any, and locate them.

If you are out after information, get it; that's your duty. At all hazards, mark the entrance and re-entrance of your own and enemy's wire. All your party should be trained to cut wire properly. Inexperienced men have gone out on patrol, incorrectly cut the wire; it has sprung, with the result that the patrol was given away, and the night's work a failure. Never let this be said of your patrol.

In returning to your own lines with information, there is a tendency in men to make a bolt for it. Do not allow such a thing. It is a dangerous practice, and, may result in casualties and the loss of information. Be as cautious coming in as you were going out. This is very important.

Upon returning to your trench each man should write a report on the night's work. Never under any circumstances talk about what you have seen to each other until you have each written a clear report on what you think you saw. It is so easy for a man to change his ideas on information received from another. This is a great mistake, for the other might not be right.

The result might be disaster, and perhaps the unnecessary loss of many men's lives.

Remember, when crawling, that if after you have gone about ten yards, and you stop to listen, you hear an enemy patrol on the move, the nearer you get to them, before they find you, the less chance they have of bombing you, for the effect of the bomb would be as disastrous to them as it would be to you.

It is not wise when on information patrols to take any equipment whatever, not even a rifle. The revolver is a very good emergency weapon if you know how to use it, and have a couple of Mills bombs in each pocket as a last resource. Leave steel helmets and respirators behind and, of course, all papers and documents. A rifle is cumbersome; it is information you want, and not an engagement with the enemy. A properly trained Scout-Sniper is safer than the ordinary rifleman in the trench.

If your morale is good, it will be only a few patrols before you are trying various schemes on the enemy, and be as confident of moving in No Man's Land, though under different conditions, as you would across Piccadilly Circus. If on any rough ground you trip or kick something which causes noise, lie perfectly still; it is the only way to escape observation, and it keeps the enemy in doubt as to your numbers, intentions, etc. If you lie still long enough he will give up watching and listening and you can proceed in safety.

six

USE OF THE SCOUT-SNIPER UNDER VARIOUS CONDITIONS

THE Officer i/c Scout-Snipers should have his men thoroughly trained for all contingencies. He never knows at what time the forces in the field will break into open warfare. When they do the scouting ability of his men will come well to the fore. Here is where observation, stalking, use of cover, etc., is essential. Supposing the troops now were in open warfare, what would be the use of the Scout-Sniper ?

Now, let us take troops at the halt first. You know by your infantry training that when a body of troops is halted, there must be a body of troops detailed to protect them; therefore a picquet line or line of resistance is planned and manned ; from this line sentry groups are put out, and ahead of these groups your work comes in. You will patrol the country in pairs to a distance of one to two miles in depth—more, if necessary—and gather any information that will be of any use to the picquet line. You must let your picquet line and your sentry groups know that you are out, or trouble will ensue, just as when a patrol is out from a trench. It is important that all roads be carefully watched, and that no enemy Scouts see you, so practise the use of cover at home before going abroad. It is essential and necessary at all times.

From this we continue to troops on the move. Picture to yourself a battalion or brigade moving up a road. The protective force here consists of the advance guard, consisting of the main and van guards. Ahead of these move the Scouts ; their business is to search everything that might hide the enemy. They will look for machine guns, road blocks, artillery, infantry, or anything likely to cause trouble. Houses, strawstacks, trees, and everything must be thoroughly searched. Even if the houses, strawstacks, etc., do look deserted, fail not in your duty by not searching; never allow yourself such a blunder. A case happened in France when a battalion was on the move in the earlier part of the last war. The Scouts passed an old house which was about twenty yards back from the right of the road. To all appearances it was deserted. The blinds were drawn, there was no smoke from the chimney, and it looked desolate. They sent back no report, not having searched it. The result was that the cyclists and the advance guards were allowed to pass, and when the main body came up machine guns played havoc with them through the dominating windows and blinds. It was too late to remedy the mistake; the carelessness of a few caused the lives of many to be lost. These things will happen unless men are chosen who can use their brains and have a keen sense of duty. Scouting and Sniping are now of such importance that the Officer i/c Scout-Snipers should have no difficulty in getting the very best men to train. That Officer is responsible for training them thoroughly. Officers and N.C.Os. are now being trained very thoroughly as instructors. There should be no excuse for careless work amongst Scout-Snipers.

Never lose track of the body that you are screening; this might prove disastrous, for if you lose time in finding them, likely by the time you do the Commander will not have time to deploy before the enemy is upon him.

Scouts will be necessary for the flanks, as well as in advance of the advance guard. You never know at what time the enemy will attempt to surprise you by moving around your flank and cutting you off in the rear.

Scouts should be used when troops are in extended order and about to engage the enemy. They should be well pushed out with relays back to the front line, in

PLATE III.

THE WRONG WAY TO SHOOT THROUGH A WINDOW

order that any message may be got back quickly, without allowing the Scout to leave his position for any length of time. The first Scout, after returning to No. 1 relay with his message, can return immediately to his position and carry on with his observations, while No. 2 relay takes it back to No. 3, and so on.

Let us take the same men being used as Snipers in the advance guard. They would work with the vanguard and use their intelligence as to action as much as possible. If coming in touch with the enemy, those who are not on Scouting duty would push forward and take up high positions, and by accurate shooting would worry the enemy, particularly trying to pick off leaders, while the Commander is making plans of attack.

In the advance from the trench the Sniper is exceedingly useful, and his work considerable. As a rule, the infantry are held up at about the third or fourth line ; this is the time the Snipers should be under the direct control of the Sniping Officer, who brings them up with the support, and puts them on the track of the machine guns which are holding back our infantry. They can be the cause of many casualties amongst the panic-stricken enemy retreaters also. Counter-Sniper work is also very important.

Snipers are highly skilled and difficult to replace. On no account should they be sent over with the first wave of attack.

The part played by the Sniper in the rear-guard is also of extreme importance. The object of a rear-guard is to delay the enemy, and, if possible, make him deploy, so as to allow one's main force to gain a place of safety. With this object in view the guarding force should be as small as possible. They are stretched out to make the enemy believe that there are greater numbers. Now the intelligence, quick wit, and originality of the Sniper are of supreme use in such a case. The game is to make each shot tell ; the more casualties you inflict, the lower their morale becomes, and time is wasted. There are many things a Sniper can do in a rear-guard if he will only use his brains, and properly trained men will never fail their leaders.

The above are various uses, apart from trench warfare, of the Scout-Sniper. What these men should know will be explained in the following pages. Each and every Scout-Sniper should be trained alike in order that one is capable of doing the work of the other.

seven

SNIPERS IN ATTACK AND DEFENCE

[*This chapter refers mostly to trench warfare of the kind experienced in 1914-1918. Much of the information, particularly in the later part of the chapter, is just sound common sense applicable to any kind of Scouting or Sniping.*—EDITOR.]

IF an attack is to be made, it will be well to have additional Sniping Posts if advanced posts are used, so that the Rovers can be on the spot, and, if possible, make posts for the Snipers belonging to the support company. They can cover the advance of the troops from the rear by quick and accurate fire, and, as likely as not, as before stated, put the enemy's machine guns out of action. If the enemy does not know from whence the fire comes, serious damage can be inflicted.

If there is a depression in front of you, and your troops advance, you can stay where you are giving protective fire, in case the enemy wire is to be cut. You must under no consideration get out of your advanced post when under observation by the enemy. Only when our men are entering the enemy's trenches do you make haste and move forward, and take up better positions, so that your infantry will further benefit by your marksmanship. You at once begin to take notice as to location of new Sniping Posts, in or in front of the new position, and have them constructed and occupied at once.

In a general advance the O.C. Snipers should see that the whole unit benefits by your marksmanship, by properly distributing Snipers along the unit's front.

SNIPERS IN ATTACK AND DEFENCE

But Snipers should not be tied down to hard-and-fast rules in the firing line, but allowed to use their own judgment as to the position which would be of most use for the work they have to do.

Snipers should never take part as an average rifleman in a night attack, for their marksmanship is of no special value. They should be on hand to take every advantage of the result of the attack, choose their best positions and man them before dawn.

In such a case keep cool. If you get a good elevated position close up to the enemy's lines, and well concealed, you can make it very warm for their reinforcements. The enemy's bombers may also require some attending to. Don't use a position too far back, for then you will come under the enemy's artillery fire. You must keep a sharp look-out for the enemy Sniper; likely he will be doing the same thing, so put him out of action before he can do you any harm, or the men whom you are screening.

If you find that in the advance it is necessary to shoot over the heads of your own men, be very careful that you do not shoot directly over a man, for he might rise at any time. When you are in front of your own men, you should be well bullet-proof concealed from the rear, because they are not as expert marksmen as yourself.

In the advance any Sniper's equipment in the possession of the individual Sniper—that is, apart from what he needs—must be left safely behind, and the O.C. Snipers must make arrangements to have this reserve collected and guarded, and reissued later when convenient. When in the firing line with other troops, give them the benefit of your superior knowledge as to range.

At any time when you find a wounded man, fix his bayonet in his rifle, and shove the bayonet down in the ground, so that a stretcher-bearer can easily find him. You may save a man's life this way, and very much facilitate the work of the search-parties and stretcher-bearers.

In case the attack should fail, concealed as Snipers on

each company front, you would be invaluable in covering the retreat of your unit across No Man's Land back to your own trenches, doing a great deal of harm to the enemy. You will remain in your post until dark, as usual, until relieved by the night guard or listening post.

In the case of the enemy's surprise attack, pick off their men until they are close to you, then lie still and let them pass. If your post is concealed as it ought to be, it will not be discovered. It is as well to have three or four extra rations, for if the attack is successful, you will now be in the rear of their lines. If unfortunate enough to become a prisoner, play the fool and say nothing; destroy your equipment or any military information. If not found out, you may do useful work for two or three days behind their lines, as they have done behind ours; but, if possible, crawl through their lines and back at night; this has often been done. You have confidence, anyway, that your troops will counter-attack, so stick and remain cool as long as possible, or as long as you may think it advisable for your own benefit and the cause in general.

In a case of this kind it is not so hard to get back to your own lines as it may appear. Penetrating the enemy's lines from his rear is much easier than from his front. Take your time; take advantage of everything to cover your movements—of any noises going on, talking of enemy's sentries, the moon disappearing behind clouds, etc. When the star-shells go up, close your eyes, as you can see better in the dark afterwards. When you approach your own lines, take cover in a shell-hole or any good cover before you challenge a sentry, and be sure it is your own sentry that you are dealing with.

When the enemy is preparing an attack, there is generally the usual preparatory bombardment of your trenches. This bombardment will not affect you if you are properly located—ahead of our trenches for preference, or in rear of them if necessity demands. It all depends on local conditions. If our own artillery is not shelling the enemy's trenches at the same time, they will consider themselves quite safe in looking over to see the effect of their shells on our trenches. That is just what you are hoping for. You can make a good bag in this way.

If your Sniping Post is in your trenches when they are being bombarded, you can do no good there; far better seek à place of safety, as you will be a valuable man to have when the bombardment is over and the enemy launches his attack. If you leave your post in the face of an enemy attack, you must in all cases blow it up. In some places local conditions will be such that you can retire without being exposed to the fire of either side, but if you have nerve and sufficient confidence in yourself, you will do more good by sticking in your post, provided it is properly concealed. An open-top post would be hopeless. There is only one thing to do—retire—and in doing so take advantage of every available cover. You may also save yourself by having a tunnel to some other place (a few yards will do) that is more concealed, but not so good to shoot from as your regular post, having just loopholes and no exit. Now, if you are discovered, and the enemy call on you to surrender, and lots of them stand about, retire to your auxiliary post and blow up your main post, and some of the enemy may go with it. They have every reason to think that you blew yourself up also, as part of connecting tunnel will have fallen in during the explosion. If you succeed at this, you can dig yourself out at leisure or when you think best. It is a poor fox that has but one hole, and of two evils choose the least. This of course, applies to a concealed post.

In the rear-guard action all Snipers should be in the extreme rear, with the machine gunners told off for this purpose. The enemy's progress can be greatly impeded, if not entirely stopped. At any rate, the Sniper can hold them up until the troops in the rear have been able to take up an advantageous position. This work should be done under the personal supervision of the

Battalion O.C. Snipers. For observation holes from Sniping Posts or a fire trench at various angles you can often use a trench auger to advantage. It only takes a few minutes to bore through, and if results are not satisfactory, you can make another one.

In advancing into enemy's country or country where you are likely to encounter enemy patrols, Scouts, etc., you proceed very cautiously, and see without being seen; search all likely and unlikely cover as you advance, and take notice of any signs of any kind that are left behind by the enemy, and search the country well ahead of you, taking care to note the things on both your flanks, as you can often, by close observation, see objects a long way off to your right or left better than those Scouts on whose front it occurs. Glance back every little while, and picture what the country looks like behind you, so that you will be able to recognize natural features upon return.

It will be rarely wise for Scouts to fire at anybody or anything; but, being highly trained men and above the average intelligence, they should instantly consider the advantage and disadvantage resulting from such action. If a lot of firing is going on everywhere, and your own firing does not give your position away, you become a Sniper as well as a Scout for the moment; so you see the difference between this kind of work and pure and simple Sniping. When employed only in the capacity of Sniper, you must always kill an enemy on sight; when Scouting, you may do the same, but only when, considering circumstances and possible consequences, you find it wise to do so. In Scouting you will often have to beat a hasty retreat, so when passing over the country take notice of all cover in your rear, so you can return without being seen. If you meet the enemy Scouts, don't ever let them go back and report to their own lines.

The troops in your rear may have to open fire at any moment. It would be a pity if you should afterwards hear: " Well, we did not succeed because we had to waste our best opportunity waiting for the Scouts to

SNIPERS IN ATTACK AND DEFENCE

come in." It should be understood that the Scouts are quite capable of taking care of themselves ; so in this connection you always examine the ground as a matter of course. It should never be hard for you to find suitable cover on a second's notice. When the firing starts you must join in ; and if it is long-range work, the trajectory of the opposing firing lines will be in your favour, and you can do great damage to the enemy yourself at your comparatively close range, the enemy not knowing you are there. Being a good Scout, you are not very likely to be caught in this way in very close action. If so, you must use your own judgment ; keep cool, and, with your ability and education in Scouting and Sniping, you should come out all right. Always bear in mind that, whatever tactics you are using, the enemy may possibly practise the same ; don't let him be first.

eight

INTELLIGENCE

EVERY Scout-Sniper should be an Intelligence Department of his own. He will, however, be a part of the general intelligence scheme and organization. He must realize the importance of keeping everything quiet. It is a very serious offence for any man to let out any military information, but with a Scout-Sniper it is worse than a mere offence; it is a crime that should call for a very severe penalty. A Scout-Sniper should know better. There has been too much information leaking out of late; wherever you go, you hear soldiers discussing this, that, and the other thing. It is wrong—entirely wrong. Men are careless when they are on leave, either from lack of something to say or because they are interested. It was just the other day that we were going up to a certain city, and an officer was home from the front on leave. We asked him a few questions as to the conditions at the front now. We got our reply all right, and more; he became so exhilarated about some of our new weapons that he told us all about them, even to the most minute details. You see that this is a mistake, but you see when it is too late. In mixing with women, perfect strangers often, military information is confidentially given. There are women who are the greatest spies that the world has ever known; they have their contemporaries, and you don't know who they are or where they are. Remember, when it is military information, be silent when strangers are about—even if the strangers are in uniform.

Intelligence organization, if fully explained, would be more than time or pages would allow. So far as the

PLATE IV.

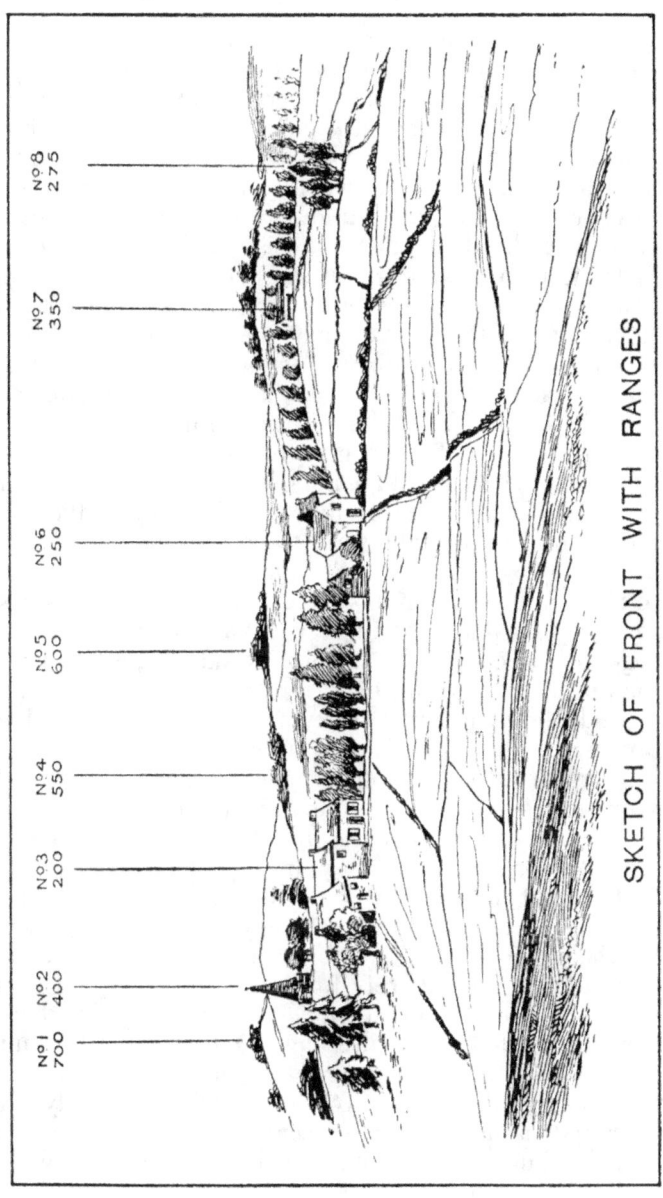

Scout-Sniper is concerned his rule is SILENCE, except to his own Intelligence Officer or others with proper authority. When on duty, whether as Sniper, Observer or Scout, let nothing escape your notice—for everything that is done there is always a reason. Keep your eyes and ears open, no matter where you are. If you suspect anyone, do not let them know it, as it will put them on their guard. The most innocent-looking and acting of those people who are out to obtain information are usually the most dangerous, for your suspicion will not likely be aroused. You may be the cause of heavy losses to life and property by too much talking. No matter who may ask, when it is about military information, say nothing. Obtain all the information you can, but give none. The enemy are constantly impressing upon their soldiers and sailors the value of silence pertaining to military matters. Never answer advertisements from people who want to write to lonely soldiers. They may be simple, kindly folk with good intentions or they may not. Things the Censor would always pass may be of value to the spy who can put two and two together. So much for the passive side of intelligence— making Silence your shield.

In dealing with prisoners your duties are clearly laid down and will be indicated later.

When a patrol or a party of the enemy approach your line to surrender voluntarily, and call upon you to come out of your trenches and take them, don't go, as it is likely a trick that they are trying to play on you. In the daytime you may be sure that it is so, for if the enemy trenches were manned they would be shot down by their own men. Call for them one at a time, and search them immediately for concealed arms. If they refuse to come in or start to run back, shoot them. Never trust to their white flag. With regard to kindness to deserters, by acting intelligently in this respect desertion from the enemy's lines will be greatly encouraged, and a genuine deserter usually brings very valuable information ; but he will only come over when he thinks it safe to do so.

INTELLIGENCE

When in pursuit of the enemy, take the greatest of precautions, and be suspicious of everything; take nothing for granted. An innocent helmet lying on the field may be a contrivance to kill by having a bomb fixed under it. Any other equipment might be similarly fixed. The Germans have been known to place live bombs under dead bodies. Guard against any buildings or bridges left intact by the enemy; likely they will be mined. Don't eat or drink anything the enemy have left behind before it has been thoroughly examined by competent authorities. Watch carefully the attitude of the local population in towns and villages, particularly if you are in the enemy's country. Watch all windmills, as they are often used for signals; also church bells, and clocks on towers. Local inns, barbers' shops, blacksmiths' shops are the usual places for information.

Small children hear and see things, and will usually tell you the truth when their parents won't. Be friendly with the children and seek all information from them possible. When you wish to obtain information do not ask point-blank questions, but go indirectly to work, and then put two and two together. When asking questions, appear disinterested; by this means you can obtain a great deal of information without the individual's knowledge that that is what you are after, and information received in this way is much more valuable than otherwise.

nine

NOTES ON THE COLLECTION AND TRANSMISSION OF INTELLIGENCE BY TROOPS

EARLY and complete information regarding any enemy's units with which our troops may be engaged is required at all times by the Intelligence Department. This information is obtained by contact with the enemy, and by searching the enemy's dead, wounded, prisoners, and deserters.

You would look for the following identification marks on the German soldiers :—

1. The identity disc, which is hung around the neck.
2. The pay-book, known in German as the *Soldbuch*, is kept in the tunic pocket.
3. The shoulder-strap, which is marked with a number or monogram. Pay special attention to the colour of these marks, as well as the piping which surrounds it. If you forward the strap in question, state whether taken from the tunic or greatcoat.
4. The markings found on arms, clothing, etc., are inside the flap of the cartridge pouch, on the bayonet near the belt, on the back of the tunic lining, and inside the cap or helmet. The colour of pipings, bands, etc., on the caps. You will find that the band is often covered by a strip of grey cloth, and on the tunics, on the collar and cuffs. Maps, letters, notebooks, orders, etc., are usually found in the skirt pocket at the back of the tunic.

The Procedure in Case of Enemy's Dead.—You will forward to proper authorities, together with a statement as to the place where they were found, all identity discs, papers and pay-books.

The Procedure in Case of Prisoners.—They are to be searched at the earliest opportunity after capture, in order to prevent them from destroying any letters or orders which may be in their possession. They should not be questioned by all and sundry. The Patrol Leader should ask the prisoner his rank, name and number. These should be noted. If in his excitement the prisoner volunteers information, this should be carefully recorded. Remember, however, that a prisoner may " dry-up " at the sight of a notebook. The prisoners will then be sent to Headquarters with the least possible delay. Lightly wounded prisoners fit to be interrogated should accompany the unwounded. The identity discs will be retained by all prisoners, but papers and pay-books will be transmitted together with, but separately from, owners.

The fuses of exploded shells furnish valuable information. When found, they should be sent to the nearest battery through Intelligence Officer, together with an accurate statement as to the exact location in which they were found.

Examine all billets and bivouacs deserted by the enemy ; likely you will find identification marks, such as chalk marks on the doors. These marks are to be copied and forwarded to the proper authorities.

Captured or abandoned vehicles or horses should be examined for markings, etc.

ten

REPORT WRITING

THE writing of reports is a very important part of the work of the Scout-Sniper. In reporting, like everything else, there is a right way and a wrong way. So many men have obtained valuable information, but have made it useless by not timing or dating it, or neglecting to state the place of observation or the exact locality of the subject of the report.

In describing the following objects, note the kind of information wanted :—

Rivers.—Width, depth, fordable, banks, bottom, tidal, whether liable to flood.

Bridges.—Material, size, number of arches, height, width of road, height of arch to level of water, etc.

Reports should be concise and should always follow the official form laid down. Generally the proper form is taught to Snipers and Scouts and must be used. Official abbreviations should always be used when they are known, but the report writer must never invent abbreviations.

In dating, use the official method—*i.e.*, 2 Apr. 40.

In timing use twenty-four hour clock, and always use all four figures. Thus :

      0005 = five minutes after midnight.
      0200 = 2 a.m.
      1700 = 5 p.m.

Bearings should always be grid bearings when using a gridded map; otherwise, magnetic bearings will be given.

REPORT WRITING

Block capital letters will always be used for code groups, place names, points of compass and for the word NOT.

Railways.—Gauge, cutting, embankment, bridges, double or single track, etc.

Buildings.—Material, size, cellars, good field of fire, etc.

Woods.—Whether passable to all arms.

Obstacles.—Description of same.

Roads.—Width, class, metalled or otherwise, hedges, ditches, etc.

The battalion that the reporter is writing from or to should never be stated in full, but the code of the same be used—*e.g.* :

From : 21356 Sniper T. BROWN.
 LOB (this representing the unit).

To : O.C. Snipers (or I.O.).
 LOB.

In case of capture, the enemy will not know who the troops opposite are.

I draw up the following report. Is it right or wrong ?

ENEMY wire.

From : Sniper T. BROWN.
 LOB.

To : O.C. Snipers.
 LOB.

LENS $\frac{1}{20000}$

No. 2 Sniping Post.

2 *Feb.* 40

Enemy wire cut 45 deg. from this post, for a width of 100 yards, and left of bearing.

Sgt. BROWN,
0430. Sniper No. 2 Post.

This is not necessarily the form you will use. You will probably find proper forms for all messages are

SNIPING, SCOUTING AND PATROLLING

laid down officially. Memorize the few you have to use, and do not depart from them.

It is a habit of some men to leave out very important items in their reports. A mistake will give no end of trouble. It is quite unnecessary on an information report to start off " Sir."

You can see the unnecessary work in the following report, the text of which expresses identically the same as above :—

ENEMY wire.

> *From :* 21356 Sniper T. BROWN.
> LOB.
>
> *To :* O.C. Snipers.
> LOB

LENS $\frac{1}{20000}$

> No. 2 Sniping Post.
>
> *February 2nd,* 1940.

Sir,

> While we were out on patrol duty to-night to get information about the enemy's wire, I found that it had been cut for a width of about 100 yards. I went back to my post, and found that the bearing to the right end of the cut was 45 deg.—that is, from No. 2 Post.
>
> Sgt. BROWN.

4.30 a.m.

Sent by Pte. T. DODDS.

If information is required about a specific subject the report should contain the information asked for, and nothing else. Any other information should be the subject of a separate report.

Daily reports should contain all information gathered during the tour of duty.

Remember, in report-writing be clear ; put everything in as concise a form as possible. All names and places will be written in block letters.

eleven

THE PRISMATIC COMPASS

THE Magnetic Compass in its simplest form consists of a magnetized needle balanced on a pivot over a card on the circumference of which are marked the "cardinal" points and the "intermediate" points.

North, South, East and West are the four cardinal points, and North-East, South-East, South-West and North-West the chief intermediate points.

The needle tends to set itself on a line running North and South, and is said to "point to the North." The North-pointing end of the needle is marked in some unmistakable way so that it cannot be confused with the South-pointing end.

The needle does not point to the North Pole or True North, but to the Magnetic Pole. This is a considerable distance from the True North Pole and is not fixed, but is slowly changing its position. In this country the Magnetic North and True North have not coincided for nearly 300 years.

In Great Britain it is near enough for marching and other simple military purposes to say that Magnetic North is 10° 45' (ten degrees forty-five minutes) West of True North. This is right enough for this year (1940). By 1946 the Magnetic Variation in this country will be about 9° West of True North.

No one knows why the Magnetic North Pole circles round the real North Pole, but we do know that in this country the two poles come together only once in about 500 years.

Besides the Magnetic Variation every compass may, and usually does, have its own small individual error.

For really accurate work this error should be known and allowed for. To test for I.C.E. (Individual Compass Error) mark off on the largest-scale map available three points visible on the ground and shown on the map and as far away as possible. Draw lines from them to the observation point on the map and calculate their magnetic bearing from the data given on the side of the map or from the latest available figures obtained from official books. One at least of the points should be in a southerly direction if the other two points are West and East of North, and *vice versa*. The average error of the readings obtained will be the I.C.E. If the average error is 2° West, then the I.C.E. of that particular instrument is 2° West of Magnetic North.

The cardinal, intermediate and " by-points " used by the mariner do not give close enough readings for accurate work, and so better compasses than the simple instrument that has been described have the card divided into a scale of 360°. On such a scale North is 0° or 360°, East is 90°, South is 180°, West is 270°. As North-East is 45°—that is, half-way between North and East—it is evident that each intermediate point will be 45° greater than the nearest cardinal point, working round the scale in a clockwise direction, that is right-handedly.

The Service Prismatic Compass is contained in a non-magnetic case about 3 inches in diameter, and differs from the simple compass in the following ways :

1. The circular card is attached to the needle, and above it, so that the card revolves with the needle.

2. It has a metal lid with a glass top on which a hair-line is engraved. This line does duty for a front sight. A slit in a metal slide on the other side of the case acts as a back-sight. By the aid of these two very exact bearings can be taken.

3. The circumference of the card is divided into degrees and half-degrees. There are two scales, both reading clockwise, but starting from diametrically

opposed zeros. The reason for the second scale is evident directly we consider the prism which is fixed below the back-sight slit. This prism magnifies the figures appearing beneath it, the eye being applied to a small hole under the slit. As the prism is at the back of the compass the reading seen through it would be 180° away from the real bearing if a reversed scale were not provided.

The Service Luminous Prismatic Compass.

4. There is a small knob on the outside rim of the box. By pressing this the card can be steadied and prevented from swinging about.

5. For night work the North-pointing end of the needle is coated with luminous paint. There is a luminous setting point on the movable glass cover and luminous marks on the inside of the lid to enable sights to be taken.

There is also in use in the Service a Liquid Prismatic Compass. In this form the card is immersed in a liquid which prevents it moving about as freely as it does in the dry type. A stop is not necessary and is not provided.

The card is transparent, and beneath the point on which the prism is focused there is a patch of luminous paint which enables the scale to be read. Generally the brief description of the Mark VIII type given above will serve for the liquid type.

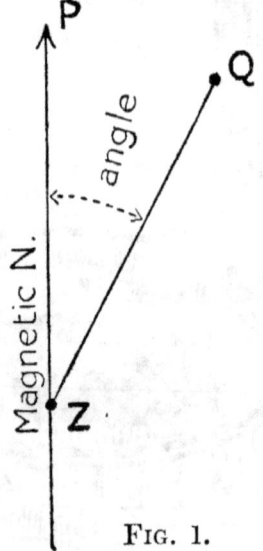

Fig. 1.

The magnetic bearing of any distant object, from the observer's position, is the direction of it, measured by the number of degrees in the angle formed by a line

THE PRISMATIC COMPASS

drawn from the observer's position to the object, and a Magnetic North and South line (in other words, a magnetic meridian) passing through observer's position. Example : If you were standing at Z (Fig. 1), and you wanted to find out bearing of Q from Z, you would have to find out what number of degrees there were in the angle formed by ZQ and ZP (a Magnetic North and South line passing through Z)—i.e., the angle PZQ.

To take a bearing on an object with the Service compass, place the left thumb through ring with left forefinger beneath the case. Hold compass perfectly level allowing needle and card to revolve freely. Raise prism level with eye. Take a sight through V-shaped slot above prism on to object, using hair-line as foresight (as in sighting a rifle). Allow card to come to rest, when the actual bearing in degrees is reflected up from rim of card by the prism.

Marching on a given magnetic bearing by day is easy. You are given a magnetic bearing, say 45°, and are told to march from your present position on that bearing for a certain distance, and that you will then come to a certain point.

With a compass with sight vanes, raise prism to eye, allow card and needle to steady, then move round slowly until figures 45° are reflected up from prism to your eye. Hold box steadily in this direction. Take a sight over V-shaped slot of prism on to horse-hair, and then into country beyond, and note some point that is in the alignment, say a house, a tree, corner of a fence, or a post. This object should be a fair distance away, say between 300 yards and 500 yards. Of course, it *may* be nearer or farther than this. March on this object, and on arrival repeat process until you have paced the whole distance you were originally told to go.

By night the luminous marks on dial and glass are used in the following way :—

The outside of box is divided into a certain number of divisions (representing degrees from 0° to 360°), the top of the compass is fitted with a revolving glass which has a luminous line painted on it. The luminous line is set to the required bearing (45° in our case) which is marked on outside of box, and all you have to do is to allow card to steady, then twist box until luminous line comes immediately over diamond-shaped North point, then the line between the luminous patches in the lid gives the direction of required advance. Set out four or five men holding up small luminous discs over their heads and align them at intervals from one another on this luminous line, then pace towards them and, as you pace up and touch the nearest man, he doubles on and takes up a position in front of farthest man and is aligned by you again in required direction, and so on.

You may be required to march on a point of the compass.

Supposing you are told, say at Aldershot, or any other town in England, to march on a point of the compass, say, for example, North-East. You see from the card that North-East is 45°, but then when told to march North-East it is intended that you should go True North-East, and your compass only gives the Magnetic North, so you must add on the variation to get required bearing. As it is 10° 45′ West or left of True North in these parts, therefore the bearing you march on will be 45°+10° 45′=56° (approx.) If the variation happened to be 5° East you would have to deduct 5° and march on a magnetic bearing of 45°−5°=40°.

To Find the Magnetic Bearing of one place from another from a Map.

(*a*) *Ordnance Map* (side edges True North and South, top edge facing the North).

Join the two places with a fine pencil line. *Through the place from which you are taking a bearing* draw another fine line parallel to the side of the map (*i.e.*, True North and South). Place your compass on the map so that centre of compass is immediately over this latter spot; allow card to steady; then, carefully

THE PRISMATIC COMPASS

raising compass a little, twist map until this last line (the True North and South one) cuts the rim of the compass card at the required degree, which is calculated according to the particular variation in that part of the country—*i.e.*, in England at 11° (approx.), or in Dublin at 17°; but in Bombay it would be 359°, as the variation there is 1° East.

Then the figures (or degrees) at the spot where the line you drew joining the two places cuts the rim of the card give the *bearing* of the place you require.

If you are using a compass with a fixed card, after needle has steadied, *before* raising compass as above, you must twist box round until figures 0° or 360° are immediately under magnetized end, and then proceed as above.

(*b*) *Military and Other Maps*, where Magnetic North is shown.

Join the two places with a pencil line through place from which you are taking a bearing, draw a line parallel to Magnetic North line on map; place compass so that its centre is over this spot and let needle steady; raise it and twist map until the last line you drew is under the North-pointing end of needle. Note where other line cuts rim of card. This gives bearing.

To Set a Map with the Compass.

On Military maps Magnetic North line is shown as well as Grid North, the Magnetic North always having an arrow head, and Grid North always a bar. Place compass over Magnetic North line so that centre of compass is immediately over the line itself. Allow needle to steady. Raise compass just clear of map gently and twist map round until Magnetic North line is immediately under North-pointing end of needle (*or 0° on card rim of revolving card*). Map is set.

The meaning of " Grid North " and True North will be described in the chapter on Map Reading, but something must here be said about converting magnetic readings to map readings.

On most maps the sides point to True North or to Grid North. It is obvious that bearings taken with a protractor on a map will not be the same as those given by the compass, for Magnetic Variation comes into the question. We have already learned that Magnetic Variation is different for different places on the surface of the earth and changes from year to year.

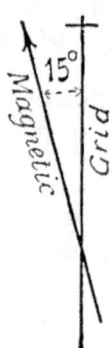

Grid North is always given as the datum line for bearings on Military maps. It is the direction given by the grid lines running up and down the map. On the side of a gridded map the Magnetic variation from Grid North is given, but in England Grid North and True North are usually taken to be the same. That they are not actually the same you can see by reference to the data printed on any Military map. In any case you will have to convert grid bearings to magnetic bearings or magnetic bearings to grid bearings.

Really this is a simple job. Look at Fig. 2. From each of the bearings given it will be seen that the grid bearing marked out by the dotted line is less than the angle of the magnetic bearings marked by the firm line by the amount of the westerly variation. The rule is therefore simple.

When variation is Westerly—

To convert Magnetic to Grid bearings subtract variation.

To convert Grid to Magnetic, add variation.

THE PRISMATIC COMPASS

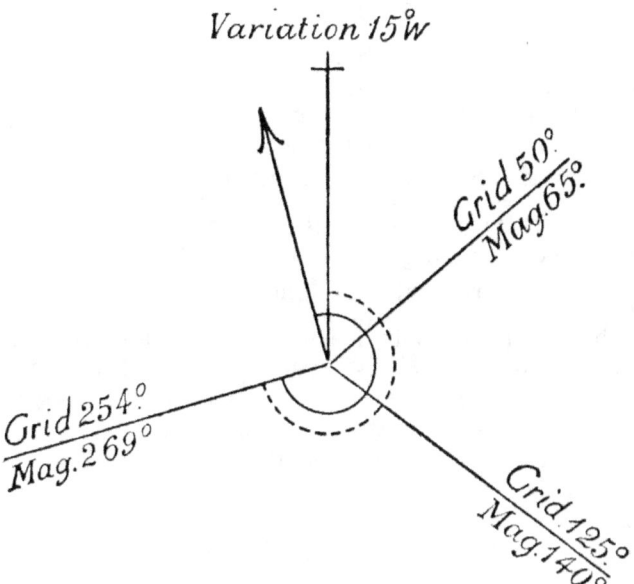

Fig. 2.

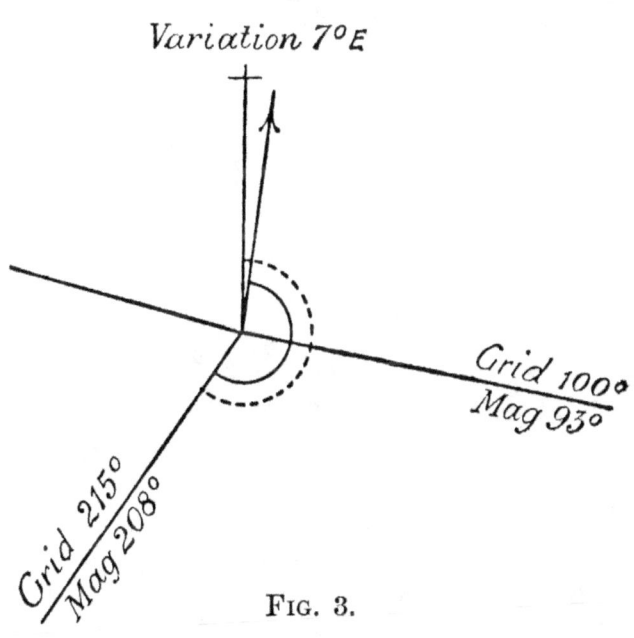

Fig. 3.

SNIPING, SCOUTING AND PATROLLING

The next case depicted (Fig. 3) shows the variation to be Easterly. Here it will be seen that the dotted line which marks out the angle of the grid bearing is greater than the magnetic bearing by the amount of the Easterly variation. The rule therefore is :—

When variations are to the East—

To convert Magnetic to Grid, add variation.

To convert Grid to Magnetic, subtract variation.

There is no need to learn the rules, however. If a simple diagram, such as those figured, is drawn everything should be quite clear and a mistake impossible. The student is advised to draw several diagrams until he becomes familiar with the idea.

twelve

THE PROTRACTOR

THE link between the compass and the map is the protractor, which provides a means of laying off or measuring bearings.

The Service instrument, " Protractor, Rectangular, 6 inch, Ivorine, ' A ' Mk. IV," is made of a celluloid or plastic material on which the protractor (angular divisions) and various scales are engraved in black. There is a long slot in the centre to enable scales adjacent to it to be placed on a map and distances read off on the map.

In laying off a bearing with the protractor, note first that the outer edge is graduated to show the degrees of the semicircle from $0°$ to $180°$, whilst inside these markings is another series showing degrees from $180°$ to $360°$. At the bottom, in the centre of the scale, is a small arrow mark which indicates the datum point from which the rays of the protractor are struck.

As a first exercise draw a thin pencil line through the point from which the bearing is to be laid off parallel to the nearest Grid North and South line. Then the arrow-head at the bottom of the scale is placed against the point, that edge of the scale being placed against the pencilled line. Any required bearings can now be laid off.

If any particular bearing is required to be known from the point already indicated as " home," then draw a thin pencil line from the place of which the bearing is required to " home " and read off the bearing of that line from the protractor scale.

SNIPING, SCOUTING AND PATROLLING

Note that for bearings less than 180° the bottom edge with the arrowhead is placed on the right of the line. When the bearing is over 180° the protractor is used on the left of the line. (See Fig. 1.)

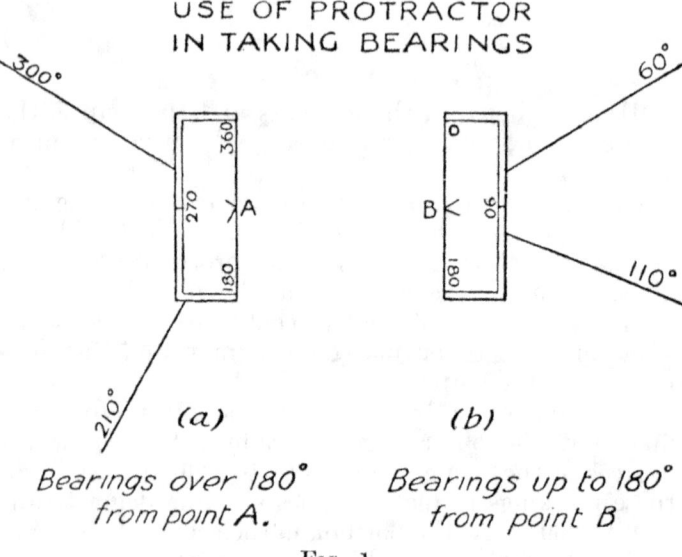

USE OF PROTRACTOR
IN TAKING BEARINGS

(a) Bearings over 180° from point A.

(b) Bearings up to 180° from point B

Fig. 1.

The two scales show the "back bearings" one of the other. For instance, if the bearing is read as 20° the back bearing is 200°. If the bearing is 310°, then the back bearing is 130°. The back bearing is always the bearing +180° for bearings less than 180° and −180° for bearings over 180°. As an example, the back bearing of the North point is either 360−180=180 (South point) or 0+180=180 (South point).

The distance scales engraved on the map are those in yards, miles and metres commonly found on maps in use in the Service. These scales will be explained in the chapter on Map Reading.

thirteen

MAP READING

IN order to be able to read a map it is necessary to know :—

1. The meaning of the conventional signs and terms used.

2. The scale of the map and how to use it.

3. The direction of True North.

4. The Magnetic Variation for that particular map and the yearly increase or decrease of that variation.

5. The method of showing differences of levels and hill features.

6. The purpose and use of the grid superimposed on Military maps.

7. The meaning of the symbols used in field sketching.

On the map itself there is usually a great deal of information. For instance, on Survey maps the most usual of the conventional signs will be found on some part or other of the sheet. A selection are figured on Plate V. It is not long before the student comes to recognize these instantly. He should make an effort to memorize all that he is likely to come across.

The scale of the map is indicated in two ways. The scale is usually shown by both methods on British maps. One way is to say that the scale is " so many inches to the mile " or " so many miles to the inch."

Familiar examples are the six inches to one mile, the one inch to one mile and the four miles to one inch.

PLATE V.

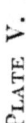

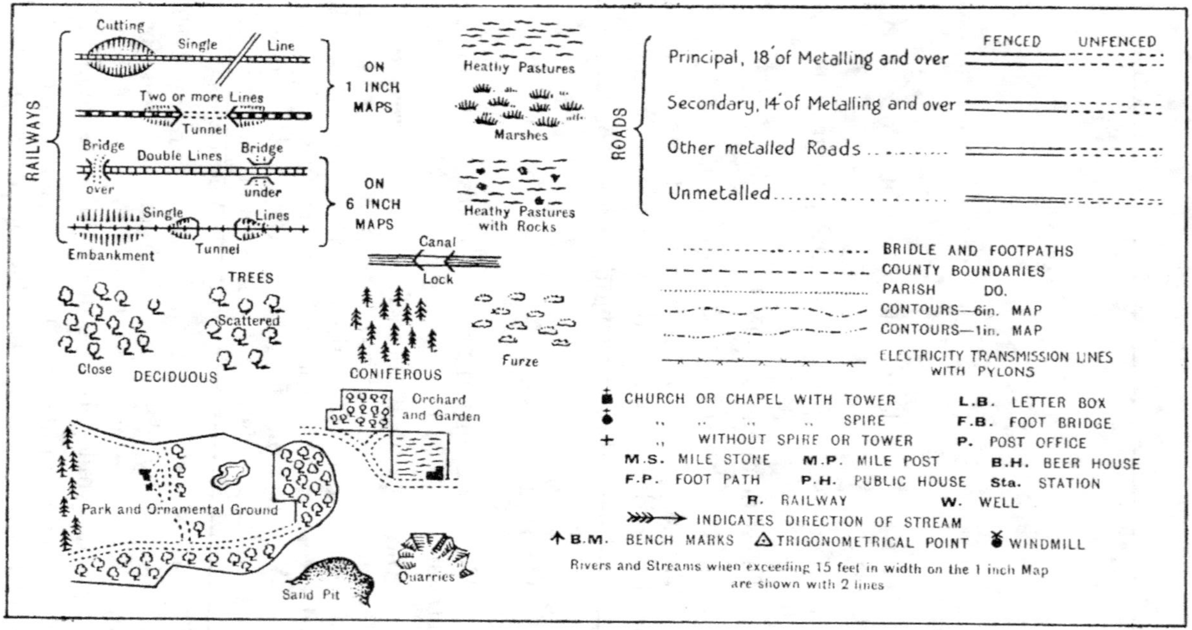

SYMBOLS USED ON THE 1 INCH AND 6 INCH ORDNANCE MAPS

The first is a large-scale map and contains a great many details. The second, familiarly known as the "one inch," is best suited to general work in this country. The detail is adequate for most ordinary purposes. It is the one the soldier will use whilst in training. The third, known as the "quarter inch" because a quarter of an inch represents one mile, is a small-scale map and will mostly be used by the higher Army formations, from Divisions upwards, where it is necessary to have a great deal of country covered by one map.

The other way of indicating scale is by means of a "representative fraction" (R.F.). For instance, the six inches to one mile map is so drawn that 6 inches on the map represents 63,360 inches of ground.

$$\frac{\text{Distance on map}}{\text{Distance on ground}} = \frac{6}{63360} = \frac{1}{10560}$$

The figure $\frac{1}{10560}$ is therefore the Representative Fraction for the six inches to one mile map. The R.F. of the one-inch map will obviously be $\frac{1}{63360}$, since there are 63,360 inches to the mile. In the case of smaller scales, four miles to one inch, for instance, multiply the number of inches in a mile by the number of miles to an inch and you have the answer. Thus :—

$$\text{R.F.} = \frac{1}{63360 \times 4} = \frac{1}{253440}$$

It will be seen that the R.F. system is really a more practical one than "inches to the mile" or "miles to the inch." It is at its best when the decimal system is used. For instance, an R.F. of 1/20,000 means that one metre on the map represents 20,000 metres on the ground, or one centimetre represents 200 metres, or one millimetre represents 20 metres.

To convert R.F. to inches to mile, or miles to inch proceed as follows :—

SNIPING, SCOUTING AND PATROLLING

R.F.$=1/25{,}000$. One inch on map represents 25,000 inches on ground. Therefore number of inches to mile $=\dfrac{63360}{25000}=2\cdot53$ inches.

Also number of miles to one inch $=\dfrac{25000}{63360}=0\cdot3945$ miles.

The Service Protractor has many useful scales for converting even-numbered foreign R.F. to yards and measuring them off on the map.

True North is always shown on a Survey and frequently on Military maps. In military work with gridded maps Grid North is always shown in its relation to Magnetic North and is always used in giving bearings. The method of converting Grid North to Magnetic North or Magnetic North to Grid North has already been explained in the chapter on the Compass.

The Magnetic Variation is shown in relation to True North on all Survey maps. If you get hold of a map on which there is no North line indicated it may be taken that the sides of the map represent the True North-South line.

Three Ways of Finding True North.

Without the use of the compass there are three ways of finding True North. The first way is by means of the sun, which at noon is due South, and North will be directly opposite. The second way is by means of the watch method. If your watch is correct, point the hour hand at the sun, bisect a line through the centre between that and twelve o'clock, and that line runs North and South, the line running away from the sun being North. (See diagram.)

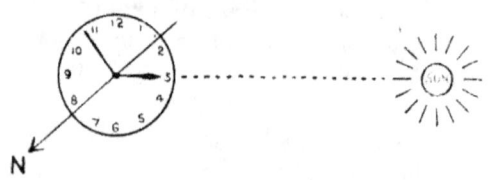

MAP READING

The third way is by means of the Pole or North Star. The diagram will show you its position with respect to the Big Bear or Dipper.

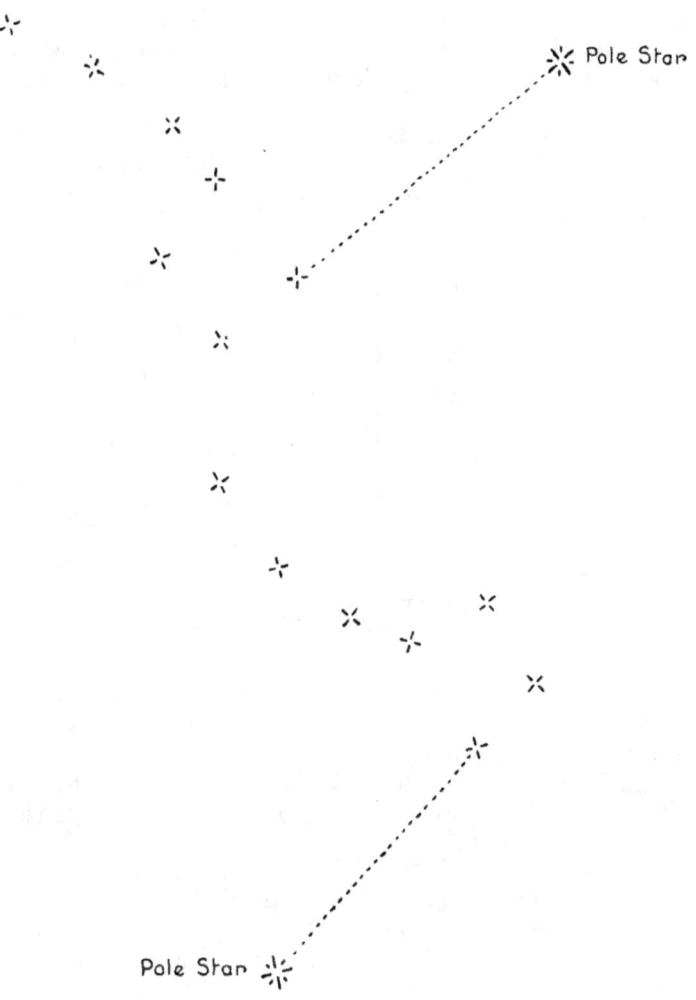

SNIPING, SCOUTING AND PATROLLING

Hill Features.

On Military maps differences of level are usually represented by "contours." These are lines usually printed in red and passing through all points in the immediate neighbourhood that are on the same level or height above sea level. If the lines are close together the slope is a steep one, and if they are wide apart the slope is a gentle one. Before you can work

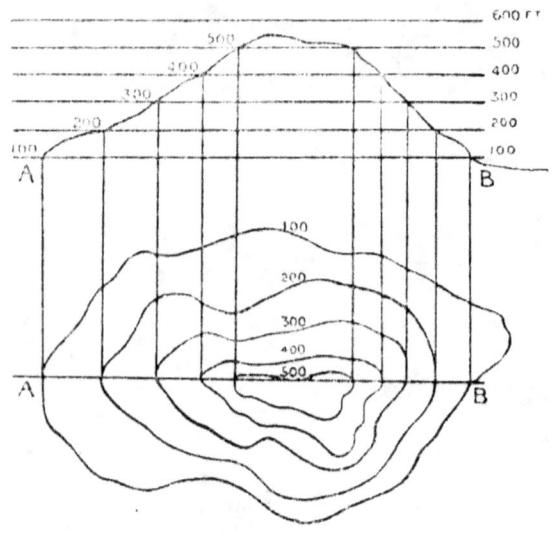

with the contour lines it is necessary to know what differences of level they represent. This information is always given somewhere on the map.

Each contour is drawn at a fixed vertical height above the other, and this fixed height is known as the "vertical interval," or V.I.

The horizontal distance between adjacent contours is the "horizontal equivalent," or H.E.

Supposing a hill is represented by five contour lines, and the V.I. is 10 feet, that hill must be 50 feet or over, but under 60 feet; for if it were 60 feet or over, and under 70 feet, it would be represented by six contours.

MAP READING

By contour lines we can always tell the slope of a hill, whether gradual or otherwise, knowing the scale of the map. Let us take five contour lines, the V.I. being 100 feet, and we want to know the slope from *A* to *B* (see diagram). Join *AB*, draw parallel lines with equal distances between to represent the V.I. (above), and make them to represent the contours, and draw perpendicular lines from *AB* up to the respective heights.

These heights are marked on the parallel, and equal distance between lines above.

In the diagram, notice that contour lines close together denote steepness, whereas contour lines farther apart are less steep. That is quite easy to see, for on one side there is 400 feet rise to a mile, and on the other 300 feet to less than half a mile. In this case the distance of the rise being found from the scale.

Co-ordinates.

The Grid System is the simplest that has yet been devised for accurately locating a point on a map so that any other person can find it.

Military maps are overprinted in purple with a grid which forms squares with sides one kilometre long. Every tenth line is thicker so as to make bigger squares each side of which is 10 kilometres long.

The small squares are numbered. To give a reference all that is necessary is to follow the direction printed on the map sheet. The only point that must be remembered is to count from left to right first. This is called the "Easting." Then count up from South to North, which gives the "Northing."

As an example suppose a D.R. at Odiham is given a message to be delivered to Eversley Cross. He has a Military map of the Aldershot district and is told that the Grid Reference of the place he has to find is 235813. He spreads his map and runs his finger along from West to East until he comes to the line running North and South marked 23. He has now determined his "Easting." He runs his finger up or down that line

SNIPING, SCOUTING AND PATROLLING

Plate VI

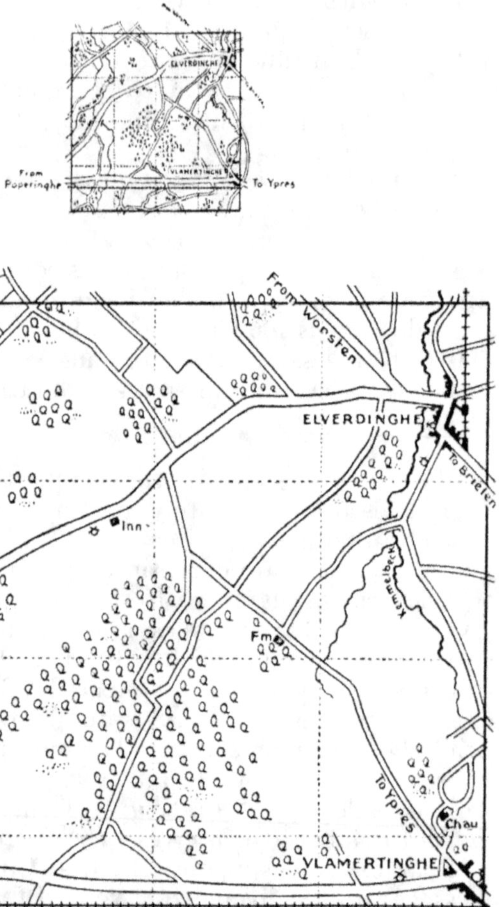

until he comes to a line running East and West numbered 81. He has now determined the position of the left-hand bottom corner of the Grid square in which his destination lies. The "Easting" was given him as 235. Therefore he moves his finger five-tenths of the line farther to the East and has established his "Easting" exactly. The "Northing" was 813. Therefore he moves his finger three-tenths of the height of the square to the North and has found the exact spot.

If you have a map of the Aldershot district do this yourself. You will find that the exact spot is Eversley Cross Church and by the sign—a simple cross—you will know that the church has neither tower nor spire.

The Grid System really is very easy indeed to understand and work with. Work out several positions on any Military gridded map you can get hold of—provided it is a fairly modern one—and you will understand very well.

But remember. "Eastings" first. "Northings" second.

You can copy, enlarge, or reduce maps by the Square System. (See diagram.) Supposing that you wanted to copy 2 square inches on a map which was 1 inch to the mile, or $\frac{1}{63360}$ four times. To do this divide your small map into squares, and draw another square four times as large, and divide into smaller squares similarly to the first; then your objects will cut the lines of the larger squares in the corresponding same place as the smaller, and your map will be enlarged.

To find Length of Sides from R.F.—A map has an R.F. of $\frac{1}{800}$; a copy is required, and on a scale of $\frac{1}{400}$. Measure the longer side of the map, and say, for instance, that it is 4 inches long. Then :—

$$\frac{1}{800} : \frac{1}{400} = 4 : x.$$

$$\frac{x}{800} = \frac{4}{400}$$

$$400x = 3200.$$

$$x = 8 \text{ inches.}$$

PLATE VII.

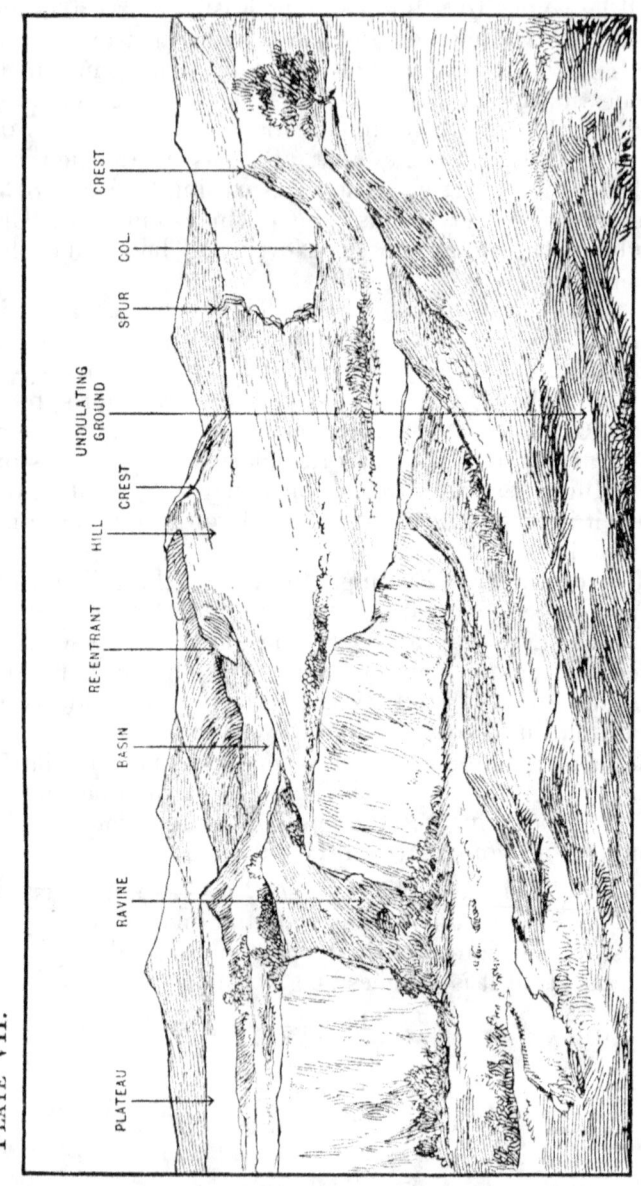

DEFINITIONS EXPLAINED

Then draw the sides and the minor squares, and continue as above. It is an advantage to further subdivide the square by diagonal lines joining the corners. The enlarged map is completed by putting in details not shown on the original small-scale map. These details can be obtained from a study of the ground.

DEFINITIONS.

HILL	Is high ground that falls away from every side.
BASIN	A small area of level ground surrounded by hills; also the district drained by a river. Think of the depression in a wash-basin.
CREST	The edge of a top of hill or a mountain, the position of which a gentle slope changes to an abrupt one. The top of a bluff or cliff.
COL	A depression between two adjacent mountains or hills, or a neck of land that connects an outlying feature with a range of mountains or hills, or with a spur.
KNOLL	A low, detached hill.
PLATEAU	A flat surface on the top of a hill; an elevated plain.
RAVINE	A narrow valley with steep sides.
SALIENT	A projection from the side of a hill or mountain, running out and down from the main feature.
RE-ENTRANT	Is where the hillside is curved inwards towards the main feature. This is always found between two salients.
TABLELAND	A high-lying, level district of country.
UNDER-FEATURE	A minor feature; an offspring of a main feature.

Plate VIII.

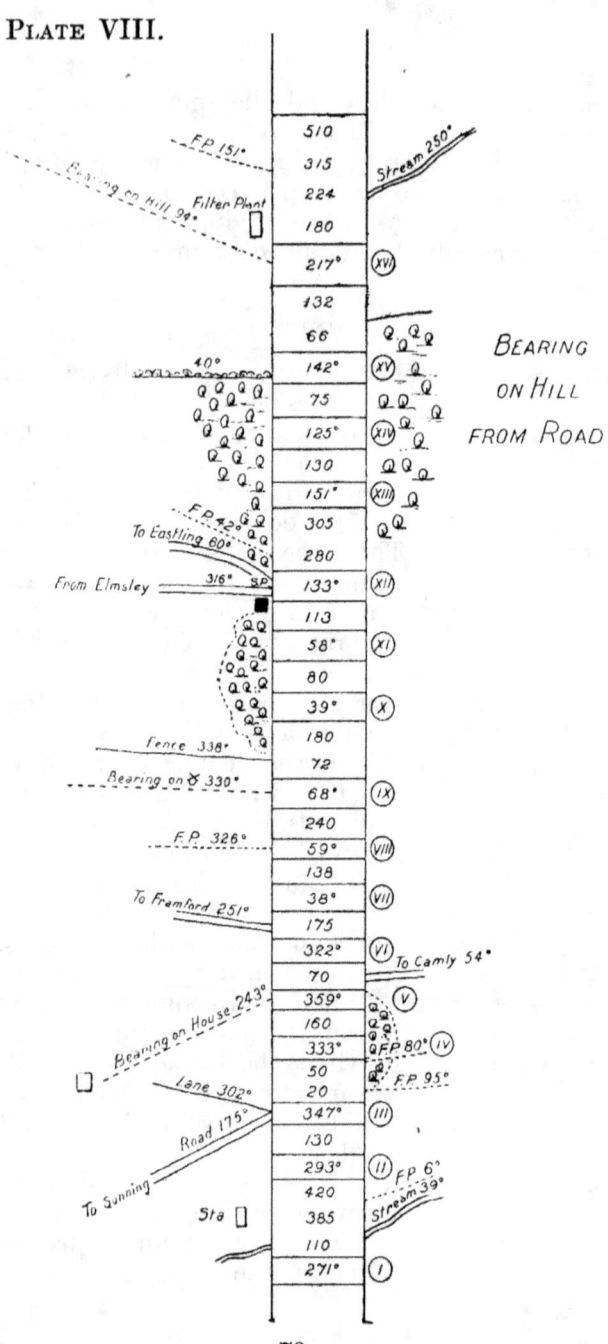

UNDULATING GROUND	Is ground consisting of alternate gentle elevations and depressions.
CONTOUR ...	An imaginary line running along the surface of the ground at the same height above mean sea-level throughout its length; each contour represents a fixed rise or fall from the ones next to it. These are expressed in feet. This fixed rise or fall is known as the V.I., or the Vertical Interval.
HORIZONTAL EQUIVALENT	Or H.E., is the horizontal distance in which a given difference of level will occur at a given degree of slope, always stated in yards; it is the distance in yards between two contours on a map.

TRAVERSING.

Traversing is a most useful method of making an approximately correct sketch of an area of country. If an enlarged map of the area is available traversing is not required as details can usually be put in by observation. Traversing is necessary when no map is available.

To traverse, a compass is essential. Traversing enables you to plot and sketch your trench system, or enables you to map any area of open country. It is done as a series of straight lines, by taking bearings in each turn. In obtaining the information, you use a chain line (see diagram), marking in all details on each side, giving bearing to each, if necessary. This work is very interesting, and with a little practice the greatest accuracy can be obtained. The procedure is as follows: In this case we start with a hill overlooking a station (I), with a circle around; it is shown in the chain line, and means the starting-point. We take a bearing on the station, and find it to be 271°, so we mark it in. Now we advance in a straight line towards the station, counting out paces, and stopping at every place that

PLATE IX.

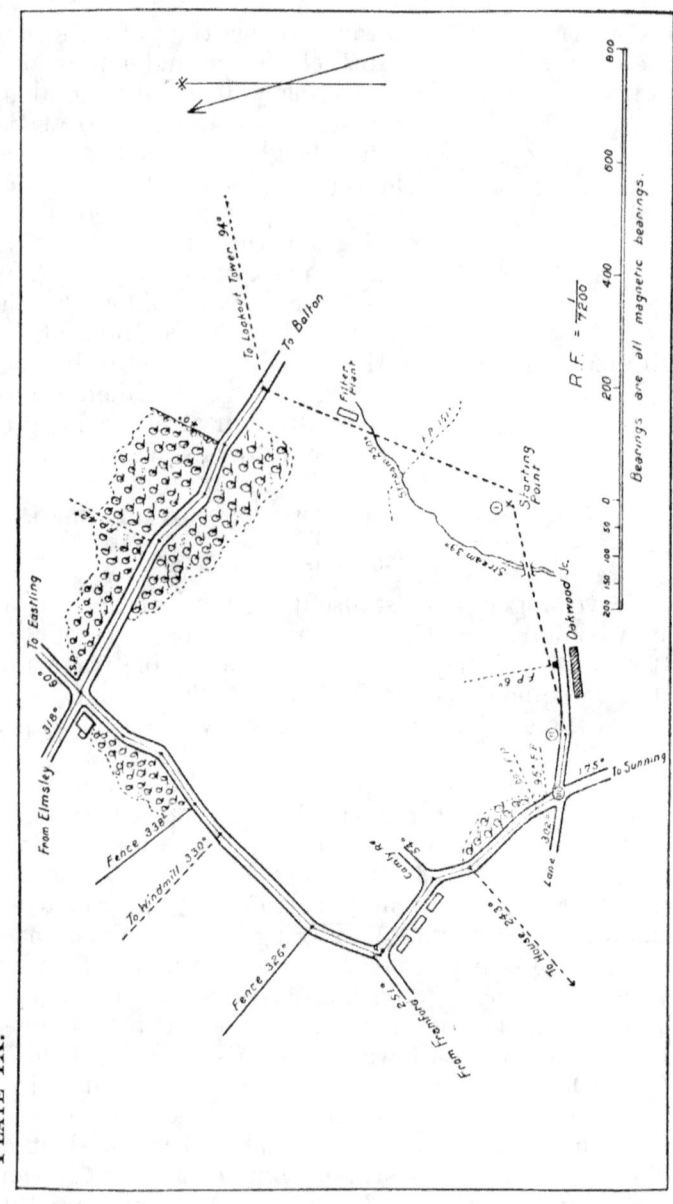

we want to mark in any information, either on the right or left of our line.

In this case we proceed 110 paces, and find a stream crossing our line, the bearing in the direction that it flows being 39°; so we fill 110 in our chain line, and the stream on the right with a bearing of the same. We now take up our counting from 110 paces, and continue on—for example, 110, 111.

Our next point is 385 paces, where we find a station on the left; we fill in the number of paces from the road the station is, etc. Here we also find a footpath going off at an angle of 106° on the right of the road; this is also filled in.

Then we continue on to 420 paces, the point to which we took our bearing. Here we draw a line across the chain line, and take a new bearing along the road which we find ourselves on; we number this point (II) and fill in the new bearing up the road as far as we can see, which is the point at which the road branches off.

We take 130 paces to the point our bearing was taken on, and find nothing of interest which we wish to mark in; so at this point we take new bearings, and call it (III), and continue on, etc. The chain line and plot in the diagram is completed. This is the work of a man who has had no previous instruction, and is accurate except for the matter of a degree or so.

To Plot.—Make a scale; in this scale it is 200 paces to the inch. Place your conventional sign for Magnetic and True North; then from your starting-point take a bearing of 271°, and the line will be a little over 2 inches long, for your first bearing consisted of 410 paces. Then fill in objects, etc., that you have taken note of, according to scale, such as stream, a little ($\frac{1}{2}$ inch) from starting-point—for it is 110 paces—and draw stream at the given bearing from your line. You continue this throughout.

Traversing is an excellent means to show quickly your exact location.

fourteen

OBSERVATION OF GROUND

OBSERVATION is one of the most important duties of the Scout-Sniper. It is an art that can be learned by all intelligent men, but professional stalkers, big-game hunters, bird watchers, poachers and country-dwellers generally have usually been at it since childhood, and the townsman finds such people hard to compete with. Observing is a fascinating game, but very exhausting to the learner. Even the highly skilled observer finds two hours of continuous watching quite enough.

Seeing things is only the beginning of the work. The observer must learn to interpret everything he sees. He must cultivate his memory so that he knows for certain whether an object is new in his sector, whether it has moved or has been moved since he last looked at that particular spot, and whether the ground surrounding it has changed in any way. In observing a trench parapet, for instance, a slight change in colour of the ground in front may mean that an enemy Sniper, clad in camouflage smock, is lying out. A very slight alteration in the contour of the ground may mean that a hide has been constructed.

The observer's best friend is the telescope. A good instrument with an object glass of not less than $1\frac{1}{2}$ inches and a magnification of 20 should be used if available, but the standard signaller's telescope with the low-power eyepiece is pretty good and may be used with confidence when there is no special issue of a better telescope to the battalion Scout-Sniper. The high-power eyepiece of the signaller's telescope should

OBSERVATION OF GROUND

only be used in very good lights and for endeavouring to identify small objects.

The telescope is made with three or four draws so that it can be closed for convenience in carrying. When not in use it should always be put into its leather case with the object-glass cap in position and the eyepiece shield closed.

In opening and closing the 'scope use a slight rotary movement. If the instrument has to be closed and put into its case while wet, take the first opportunity of drying it in a warm room with the tubes fully extended. The instrument should not be put close to a fire, but it may be dried in the warm sun. In the tropics or when the sun is very hot, drying operations should be carried out in the shade.

If the lenses become dirty they can be cleaned with a piece of chamois leather kept specially for the purpose or, in default, with a clean pocket handkerchief. In use the sunshade, which can be extended in front of the object glass, should always be used. The principal virtue of the shade is that it prevents light, particularly direct rays from the sun, from being reflected by the lens and so giving away the position of the observer.

A telescope is a remarkably easy thing to see and identify, particularly when it is waved about, and any available cover should be made use of. The successful Scout-Sniper moves his telescope very slowly when getting into position. An hour in the field searching for other students who are supposed to be hidden will quickly teach the learner how essential it is to hide his telescope as well as himself and to move it very slowly.

First lessons in observing may conveniently take the form of looking for a definite number of objects placed on the side of a hill not closer than 75 yards or farther than 150 yards away. Any military objects can be used and might be made to simulate the debris after a battle. Instructors will find it convenient to start with a small number of objects, say a dozen, making them fairly easy to find. Men more or less concealed in the same kind of country provide a useful start in progressive training which might go on to the observation of a dummy trench provided with Snipers' and observation holes, the parapet being made to look as much like the real thing as possible. In this exercise men should move about in the trench, Snipers should crawl out, and rifles, telescopes and periscopes be used.

Competition is very useful in the instruction of observers, and a list prominently displayed in the classroom, showing the points gained (one point for each object identified), is a useful incentive.

In observing for movement the telescope should be directed on to the ground on each side of the suspected spot. The sides of the eye are more sensitive than the centre, and by looking slightly away from the thing that may be moving any movement that does take place is caught by the corner of the eye, so to speak. This perception of movement in the sides of the eye is probably the remains of an even acuter full perception, for in the days of our very early ancestors the life of a man would often depend on his being able to see the slightest movement to the side.

There is no doubt at all that man has lost a good proportion of the senses of sight, hearing and smell upon which he once depended for his existence. It is probably better to put it in this way. The senses may have deteriorated a little, but the interpretation which the brain puts on the messages received from the sense organs has changed very greatly through the ages. We know, as has already been indicated, that there is a great deal of difference in the observing capabilities of the countryman and the townsman, though both, when subjected to ordinary optical and aural tests, may be equally good.

The Scout-Sniper has to restore to himself as far as possible the perceptions of his remote ancestors. No written hints will teach a man how to observe. Actual practice is necessary and, if possible, should take place under the direction of a skilled stalker. Anyone who has gone out after deer in Scotland or big game in other parts of the world knows how the professional stalker or guide teaches him what to look for and how to interpret what he sees.

At present (April, 1940) there are few men from the actual battle-front skilled in sniping and observing who can be spared to act as instructors, but there are men who had great experience during 1914-18 who are still alive and active, though now too old for actual fighting. Without doubt, they would volunteer if asked for, and should be employed. Professional hunters and stalkers from different parts of the Empire might also be employed in training.

Judging Distance.

The Sniper is seldom called upon to fire at a target at a range greater than 400 yards. Most shots will be at objects 200 yards or nearer. Very accurate distance judging up to 300 yards is therefore necessary. With training any man with sufficient brains to make a good Scout-Sniper should be able to judge the distance of an object 300 yards away to within 50 yards. That is, he should be able to say with certainty that it is more than 275 and less than 325 yards. At 100 yards he should be able to judge the distance of an object almost exactly.

Scouts should be able to judge distances with reasonable accuracy up to about 800 yards. Longer distances can be taken from the map.

Bright lights tend to overestimation and dull lights to underestimation of range. Practice at all kinds of objects, placed at known and then at unknown distances, is very good for the beginner. Generally it will be found that—

Distances are overestimated

> When men observed are kneeling or lying.
> When both background and object are of similar colour.
> On broken ground.

When looking over a valley or undulating ground.
In avenues, long streets, or ravines.
When the object lies in the shade.
When the object is viewed in mist or failing light.
When the object is only partially seen.
When the heat is rising from the ground.

Distances are underestimated

When the sun is behind the observer.
In bright or clear atmosphere.
When background and object are of different colour.
When the intervening ground is level or covered with snow.
When looking over water or a deep chasm.
When looking upward or downward.
When the object is large.

fifteen

SNIPING POSTS

SNIPERS should be thoroughly trained at choosing and digging posts. This training should be done at home or in " back areas." The training ground should be in a part of the country where the natural features vary, where there are old houses, hedges, undulating and level ground, etc. The front abroad is wide, and training should be under as many different circumstances as possible.

The ideal post is invisible from any position on its front or flanks, covers the desired frontage, and is reasonably comfortable. It should be arranged so that it can be occupied or left by a route or routes not under enemy fire or observation.

In training students may be given a line supposed to represent an enemy trench on which are marked one or two objects, fairly widely set apart, upon which fire must be possible. They will then be told, and helped, to choose likely places for the building of posts and proceed with the work. When the work is completed, students should view it from the enemy position.

Points to be Considered.

If the post is in the open, remember, do not make your trap-door too large, and do not forget to protect the sods.

Do not throw the earth all over the place, for if you do it will be readily seen by the enemy. Take a blanket or ground sheet out and lay the earth on it. Then

PLATE X.

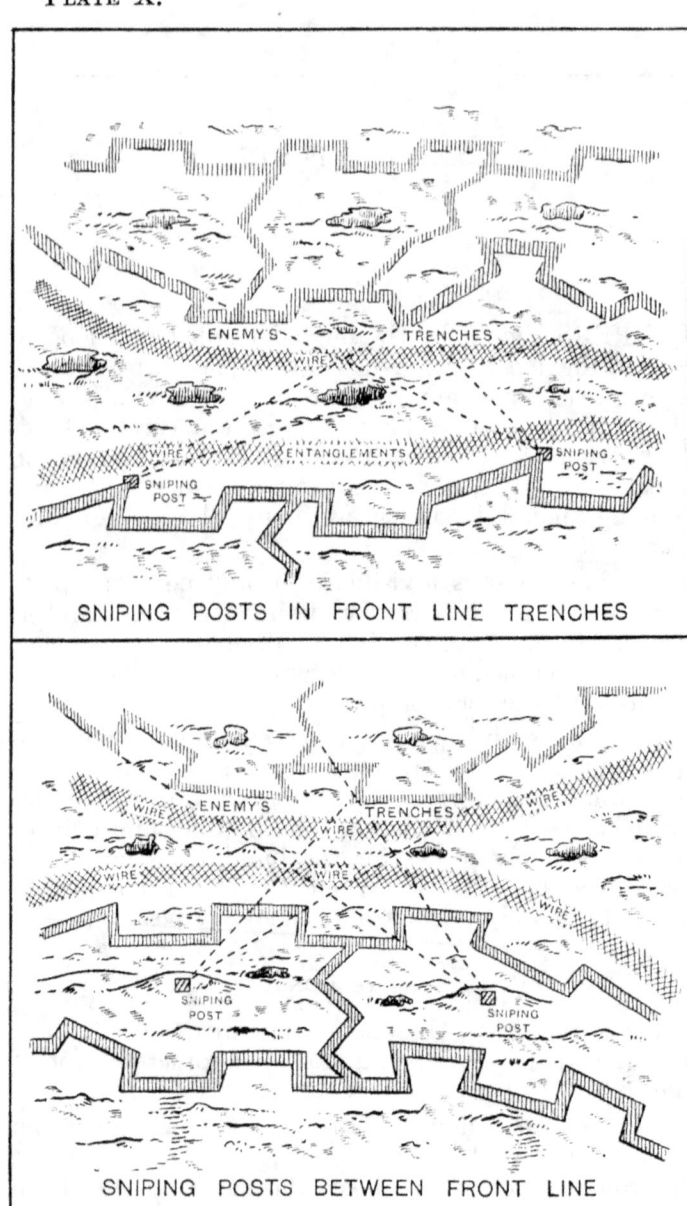

PLATE XI.

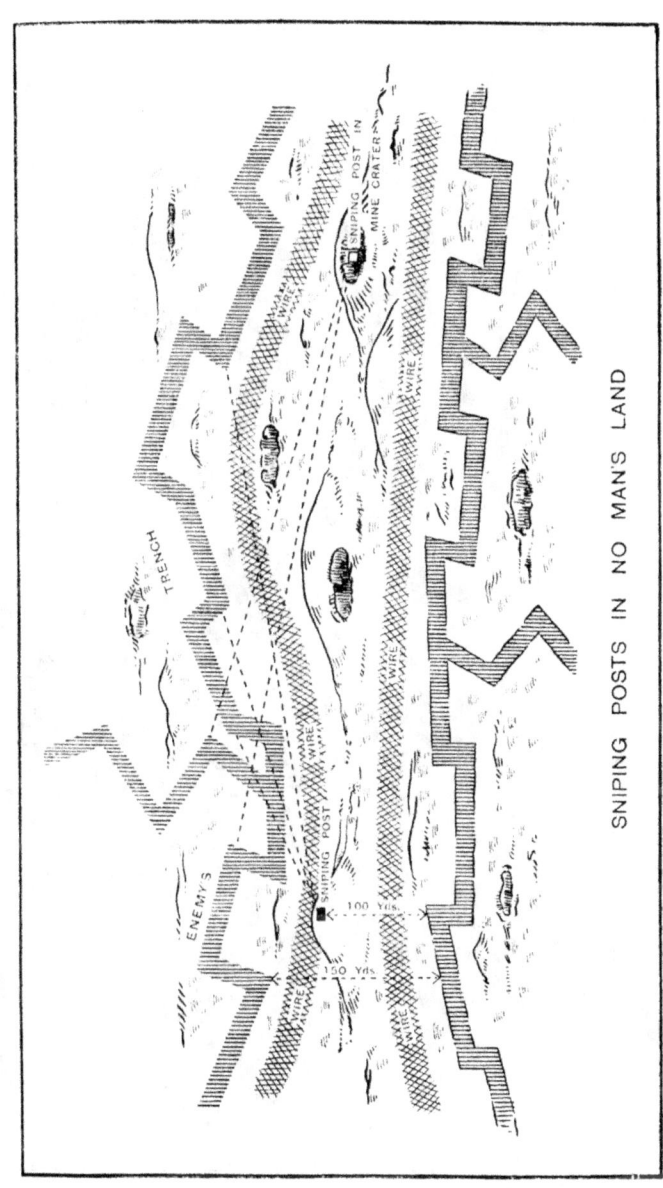

SNIPING POSTS IN NO MAN'S LAND

PLATE XII.

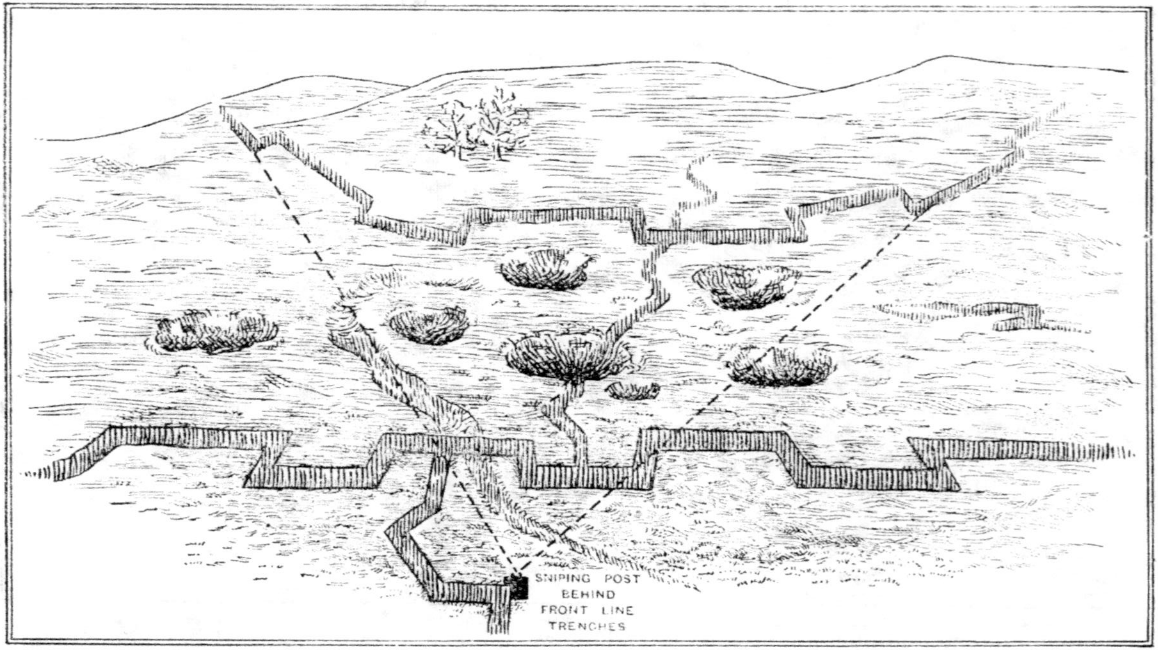

SNIPING POSTS

carry it away to some unseen spot, so that the natural features will not be disturbed.

Leave no tracks whilst building or, if you do make tracks, remove them.

The post must be dug and concealed so carefully that, when completed, a person could be three feet away and not notice it; for this reason the natural features must not be disturbed in the least. Your suspicions will be aroused if at some dawn you notice even gorse cut in No Man's Land, so will the enemy's; so take the greatest care not to disturb the natural features.

We have seen gorse placed in front of a loophole to hide the flash; the idea is all right, but if you do this you must remember that unless it is continually changed, that gorse is going to die. The colour will change, and perhaps give away your post. A fresh gorse bush, or fresh tussocks of grass, where there were none yesterday will always arouse suspicion and very likely draw machine-gun fire.

When the post is completed to the satisfaction of the men who are going to man it—this is in training—they should occupy it at once, and construct range-cards of the front as seen through the loophole. The range-card will be hung just below the loophole plate, as shown in the diagram (see page 85). Range cards, showing all prominent features, are an important part of the Sniper's and observer's equipment.

On the completion of these posts, the Instructor should go round, taking the whole of his class, and criticize the mistakes made. Now is the time for the students to ask their questions. The more questions they ask, and the better their answers, the better the class will get on. The Instructor should go through everything, leaving no point untouched.

If time is available, it is a good thing for students to have practice in making posts by night, having planned their post position, etc., by day through periscope or telescope. When men are digging at night, have half the class out in search of them. On a clear night, unless men are very quiet, pick and shovel will give them

PLATE XIII.

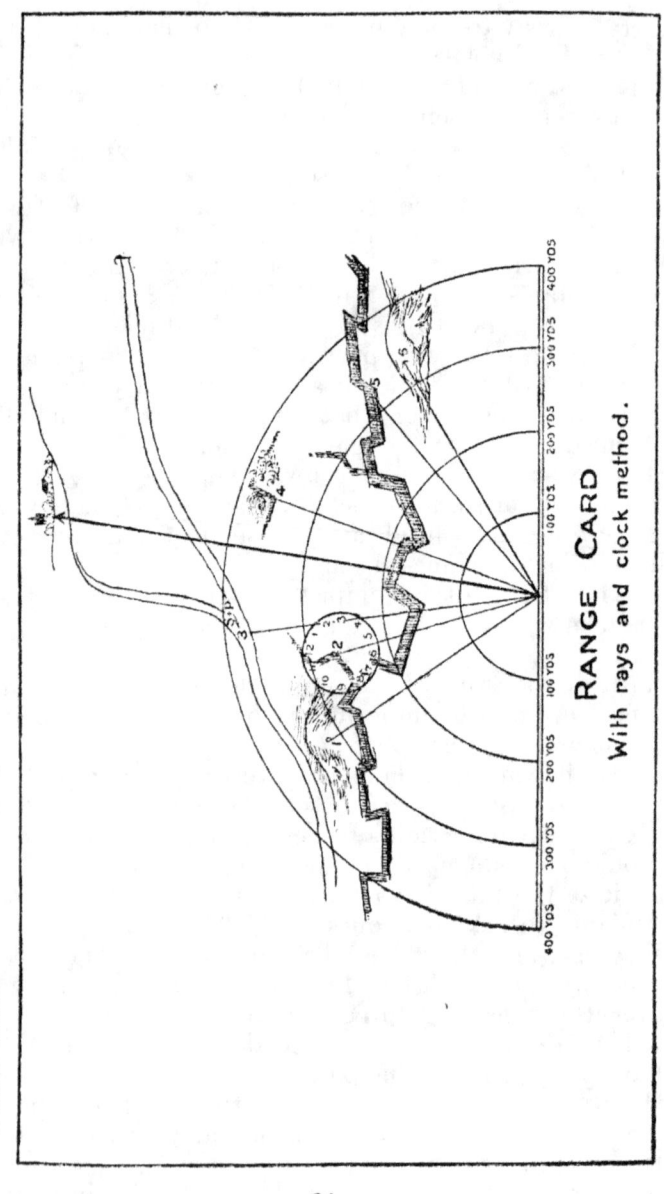

Plate XIV.

VIEW FROM LOOPHOLE – SHOWING RANGE CARD.

away. At the post, while one is digging, have one man out three or four yards on guard, to see that you are not discovered.

It has come to the notice of men occupying a Sniping Post that the effect of the Sniper's firing interferes with the work of the observer due to the discharge of the rifle. In this case it would be advantageous to have a partition between the two in order that the systematic work of the observer will not be affected in any way. A speaking tube could be arranged between the observer and Sniper so that the observer's remarks can be immediately conveyed to the firer.

If in the front line, it is sometimes well that the Sniper and observer should have separate posts about 5 or 6 feet apart, communicated between by the above method. But nothing should be done that will interfere with efficiency.

sixteen

STALKING AND THE USE OF COVER

BY making use of natural cover a good and determined Scout can usually get within a few yards of an enemy without himself being seen or heard.

The modern battle-dress is much better for stalking work than the uniform worn during the last war, but even so it is well to have a veil or mask over your face or wear a stockinette cap. You must also wear gloves; woollen khaki ones are good in the winter. In summer cover your hands with the same material as your veil, or rub them well with mud from the nearest ditch. Black ammunition boots will give you away if you are not careful. Light rubber-soled shoes are often better footwear than boots. Your rifle should be painted to suit the surrounding conditions.

When out stalking, you should leave all unessential equipment behind. Steel helmet and gas respirator should never be taken on a stalk. You may run a risk by being without them, but if you take them they are sure to give you away sooner or later.

When stalking, it is most important to watch your flanks as well as your front. It does not matter how well you are hidden from front view if you are caught from one of your flanks.

You can crawl through long grass and standing grain or corn, when there is a wind blowing, without detection. You may also be able to shoot from such cover; if so, always choose a forward natural slope.

While firing from a concealed post or natural cover, fairly close to the enemy's lines, you must work the bolt of your rifle with great care, as the sound of your bolt is easier to locate than the sound of your rifle fire ; also, in an open post or natural cover, the empty cartridge cases ejected into the air may glitter in the sun, and may be seen by a close observer. You will rarely have to fire rapid unless an attack is on, and then in the din of battle it does not matter so much. Your main point is to be able to fire one shot, or your first shot, quickly and accurately.

When there is no wind, and you creep through a field of standing grain or grass, a keen observer will soon see the unnatural movement ; and if you put your head up, it is like putting your head out of water—easily seen. You may fire from a forward slope of a grass or grain field if you are not so close to the enemy that he can see the grain moving from the blast of the discharge.

In occupying any cover, be careful of your background. Never shoot from, or occupy, the top of a large lone tree. It may in some cases be all right for observation if nobody is shooting in the neighbourhood. If you are shooting from a forest, choose the least conspicuous tree, and keep away from the first line of trees. Never shoot out of the top of a tree from the side nearest to the enemy ; the blast from the discharge of your rifle will give you away when it is calm, and the opening and closing of the branches caused by the wind will expose you to a keen observer.

Never have a wall for a close background. Also, be very careful of metalled roads for a background ; the road may give you a silhouette effect. A road is really a good place to keep away from. The enemy have also the range exactly of all roads and railways.

Look out for the enemy aircraft. Lie quite still if they are flying low.

If shooting from a room in a house, and the wall behind is not too far away from the window, make a row of small holes at a height to suit you on this inner wall, so as to get at different angles when shooting. Of course, you shoot from the other side of the wall, and in this case you can put your rifle a few inches through it. If the enemy shoot in through the window, they are not likely to hit you through any of these small holes on this inner wall. Cover up any window or door behind you. The room need not be dark, but you must have no silhouette effect by having an opening directly in your rear. You can also cut a hole in the outside wall to shoot through, but never let your rifle protrude. If even half a brick is missing in a wall, it is easily noted—at any rate, easily seen—and if any bullets are coming in from your direction the hole is sure of attention from the enemy. The range of houses, like that of roads, is known to the enemy, who will use both S.A.A. and big stuff on them fairly often, just on the chance. Therefore keep away from houses and other buildings if you have any choice.

In using telescopes or field-glasses, be careful that the sun does not make a heliograph of your lens and give your position away. The same care must be taken in using your watch and compass, even your pocket or Service issue knife, or any other bright article. A ring on your finger might give you away also.

You must learn to move slowly and evenly over comparatively open ground, so that your movements cannot be detected.

The art of stalking consists in making use of every possible bit of cover. Crawl if you have to, but do not crawl when the cover would hide a six-foot man. A good stalker will often go a mile to reach a point four hundred yards to his front. He will consider the lie of the land with regard to folds in the ground, hollows, herbage and bushes. If he finds he cannot reach his destination he will go back to some convenient point and make another cast. He is a man of infinite patience. He stops at frequent intervals, listens, and has a cautious look round, not over, cover. He watches his flanks and rear as well as his front. His object is to get where he wants to go without being seen or heard. To this end he neglects no precaution. When he crawls he is absolutely flat; if he is crawling it is because he

SNIPING, SCOUTING AND PATROLLING

is in danger. Therefore he does not crawl far without stopping to consider. Crawling flat means pulling himself along with his hands, or with one hand if he is carrying a rifle or revolver.

Note for Instructors.—In connection with this work, all Scout-Snipers, when in billets behind the lines or before going to the front, should constantly practise themselves in this work. One party should work against another; get around and through each other's lines without being seen or heard. At home we should use blank ammunition for this work. Tell students not to shoot unless they see somebody with a reasonable chance to make a good hit, and it is surprising how little ammunition will be used. In France it may not be practicable while practising in rear of lines to fire blank ammunition, but it would work out nicely in the following manner:—

If you see a man, cover him with your rifle—be careful that it is not loaded—and walk straight to where he is. If both see each other, the one that gets up first or challenges the other first wins, and the loser will come over and work with the other party. The two go far enough back to get cover to deploy, so they will not be seen in the same place again, and then proceed cautiously forward and continue the practice. In the case of a mistaken declaration, the man who makes the mistake is a prisoner to his enemy. It is surprising how quickly students progress by this method if it can be used. They begin with watching their front only, but in a short time they watch in every direction. In fact, after a few hours they have acquired at least the elements of the great stalking game.

seventeen

THE LOCATION AND CONSTRUCTION OF SNIPERS' AND OBSERVERS' POSTS

NEITHER as to the methods of construction, nor as to the location of artificial cover for Sniping Posts can any hard-and-fast rules be laid down, as there is such a vast difference with regard to local conditions. What would be good in one case would be useless in another.

Before you can construct an artificial Sniping Post you must have a certain amount of natural cover to begin with, unless it is in your fire trenches, but even in that case there must be no fresh visible work done on the forward side of the fire trench.

When operating from the fire trenches or parados the following will be of great value :—Outer face of fire trench should be as irregular as possible, both as regards forward front slope and top of parapet. Have strewn about the forward slope of trench pieces of board, sticks, small brush, empty bottles, empty bully beef and jam tins (these should be first cleaned for sanitary reasons)—in fact, all other odds and ends that you have no further use for. You can tunnel inside of trench for concealed loopholes. An old cap over your loophole on outside of parapet, with a small hole in top for looking through, will make a good Observation Post. You can make a nice place to shoot from by placing a bully-beef tin or jam tin over your loophole. To make this successfully you must have several other tins lying around ; this kind of work must be done gradually.

SNIPING, SCOUTING AND PATROLLING

You can also, on any forward or grassy slope on parapet or elsewhere, construct a wooden trap-door or hinge; on this door place about two inches of green sod, same as surroundings. Sods, or pieces of sandbags, can be fastened to trap-door with long nails. Such a door should be opened very gradually for shooting and observing. You can also line half a sandbag with wire-screening to hold it in shape and place it between the other sandbags. If well done, it will defy all detection. Very small holes will do for observation.

Plate XV.

INTERIOR OF LYING SNIPING POST (REAR VIEW)

CONSTRUCTION OF SNIPERS' AND OBSERVERS' POSTS

The enemy practise all sorts of tricks, and you must be equally cunning. He has all kinds of rubbish on the forward slopes of his trenches—old beams, brick, pieces of furniture, old mattresses, etc. All that rubbish is not there for fun or from carelessness, but it is to shoot and observe from. You must shoot into all these things from time to time, but be very careful that the enemy does not do it to you first.

The best place for concealed Sniping Posts is in front of our fire trenches, if local conditions will allow as to space between ours and enemy trenches. Where trenches are close together and the ground in the rear is too low, the last and only choice is in our own front line.

A Sniping Post should not be located in the trenches if there is any other possible position. It is better to locate the post ten yards in front of our own trenches, and to tunnel to it. When there is shooting going on the enemy will look for it coming from our own trench. If the trenches are only 50 or 60 yards apart, a post can still be constructed a few yards in front, with small loopholes out through a bunch of rank weeds or a small forward slope, and be effectively concealed even at this short range.

If the water is not too near the surface to make tunnelling possible, it can still be done by digging pretty well down. Make a kind of shallow well inside your trench, and start a few feet from the bottom and tunnel on an upward slope, drain back into the well and pump out from there. While this can be done all right, there is no use doing it if other suitable positions requiring less work and maintenance can be found. You also at this short range take a chance of being undermined and being blown up. But the greatest objection to a post so close to the enemy's lines is that you only cover a narrow front of his trenches, unless you have loopholes at several different angles. If you can occupy a higher position you can look into and behind the enemy lines.

The parados should also be considered in locating a

PLATE XVI.

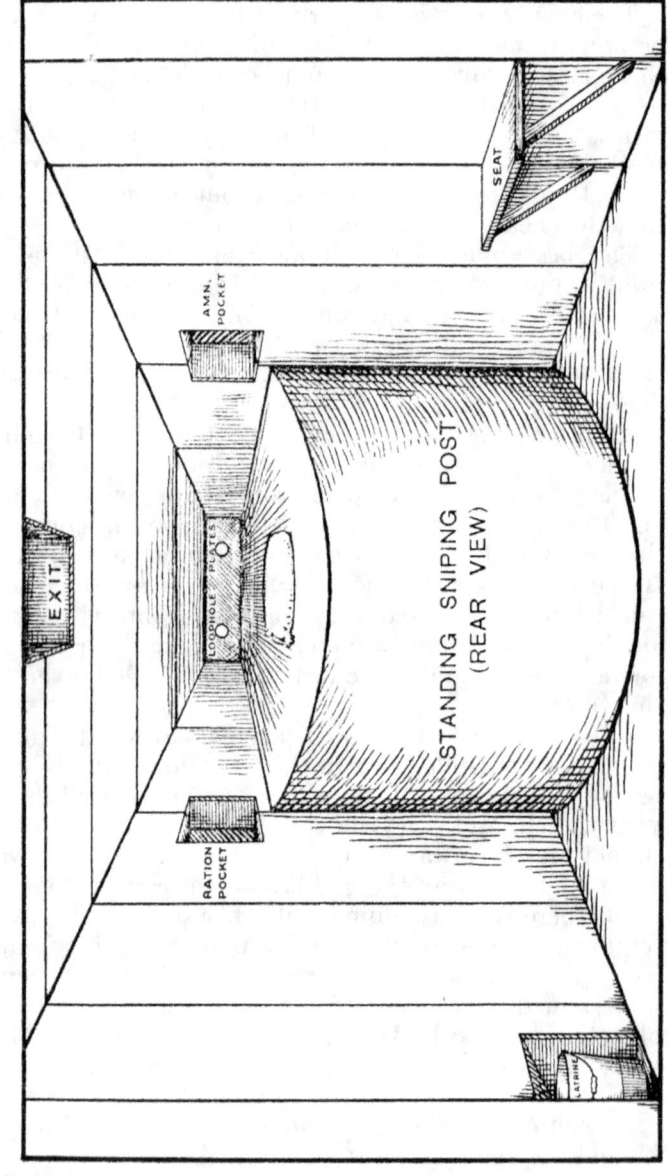

CONSTRUCTION OF SNIPERS' AND OBSERVERS' POSTS

Sniping Post, as it gives a more commanding position than in the parapet. But the parados is not a really good position. In places where trenches are only 20 and 30 yards apart it is not practicable to fire from your own front trenches, but in the rear, if the ground slope makes it possible, good positions can usually be found.

In operating a very advanced Sniping Post, where trenches are, say, 400 or 500 yards or more apart, care should be taken not to walk enough in one place to make a path that will be found by the enemy patrols. The result of this might be the mining of your post by the enemy, or in any case it would be the means of locating your Sniping Post.

It would be a great advantage in such an advanced post to have a telephone, as it would be unlikely that any other way of communication would be possible in daytime. We have ourselves occupied such posts and keenly felt the need of a phone, mostly for the benefit of our artillery, and also the O.C. Snipers, or something you want the Rovers to look up, or decoys that you might want displayed at a certain point.

When the trenches are far apart the enemy expose themselves a great deal more than when the opposite is the case. Hence the great advantage of being close at hand, where you can shoot quickly and accurately.

The forward slope of a hill is the best place for a Sniping Post. If the hill is narrow, tunnel through from the rear ; if not, dig down from the top and tunnel out to the front. If you are not under some trees or bushes, you must properly conceal the opening to correspond with the natural features. In some cases it may be necessary to drain seepage into the corner of the dugout and have a hand-pump, with a rubber-canvas hose, to take the water some little distance away. The hose must, of course, be concealed.

If there is any rise in the ground at all such a post can be made quite comfortable. In all these kinds of dugouts you dig as near to the surface as possible when you are tunnelling out for your loophole, and put

PLATE XVII.

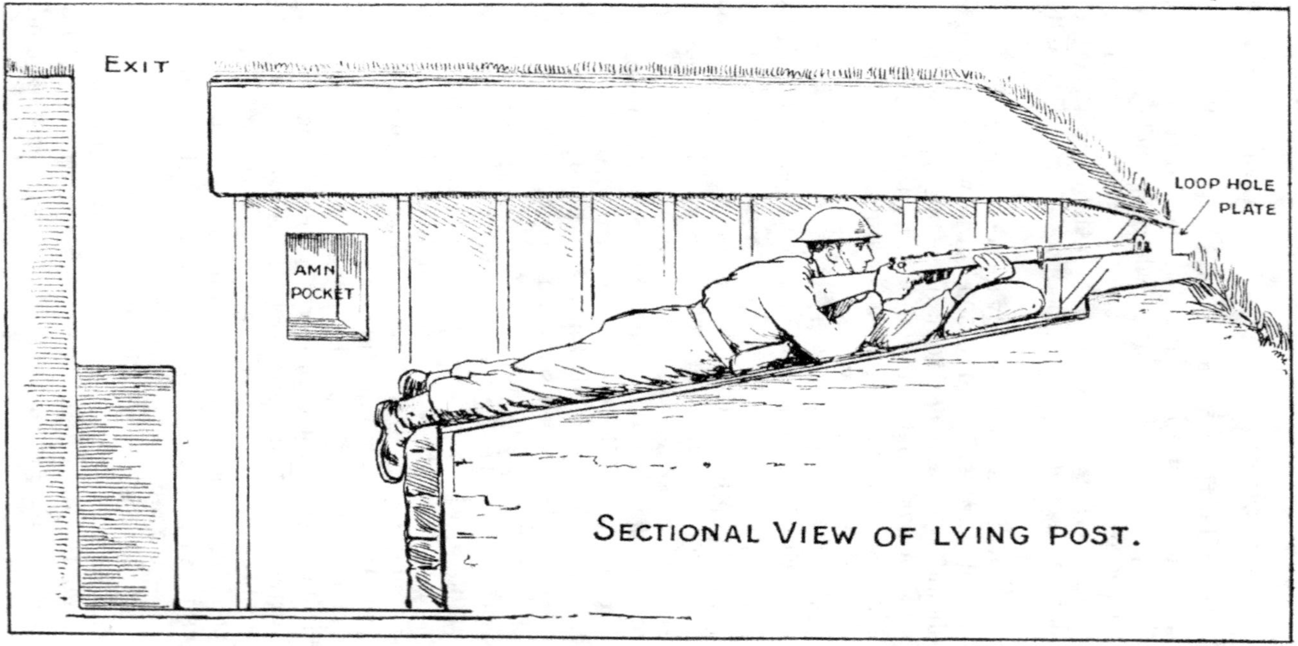

SECTIONAL VIEW OF LYING POST.

LOOP HOLE PLATE

AMN POCKET

EXIT

CONSTRUCTION OF SNIPERS' AND OBSERVERS' POSTS

two steel plates, one for each occupant, tilting them inwards at the top, and support them so that the slope of plate will be parallel to the slope of the ground, and thereby get a wider angle of fire.

You dig out to the surface only the width and length of the two plates, then widen out from there back, so that you can move your body and shoot at a wide angle. You must always locate from the front the place you want your loophole. You do this by going out at night and pushing a stick inwards. Mark the number of inches on it, and when you find it from the inside, while digging, you know exactly how far to the surface. If you don't do this, you might land right on a tuft of grass that you need to conceal your loophole. Also have the shelf that you are lying on considerably lower at your feet for comfort. Behind this point you dig down far enough to stand upright with ease. You can make yourself quite comfortable. The earth roof must be well timbered, to stand up well, as you might be there for a long time.

If the water is not too close to the surface construct your post so that you can stand up to shoot. When you shoot through a loophole you can get the different angles quicker when on your feet than when lying down.

You can also tunnel into a patch of weeds from a disused trench, and river and creek bank. From a forward slope, covered with weeds, small bushes, etc., you can often have an open-top post. If so you should have dug in the side a space that will shelter one at a time to lie down.

It must be borne in mind that except in a few favourable circumstances all these differently located dug-outs, with loopholes to shoot from, must be constructed at night. Only when you put such a dug-out in your parapet or parados can you work at it in the daytime.

A post in the parapet should be provided with a double loophole plate, the outer one being stationary, the inner one, being about two feet behind the front

Plate XVIII.

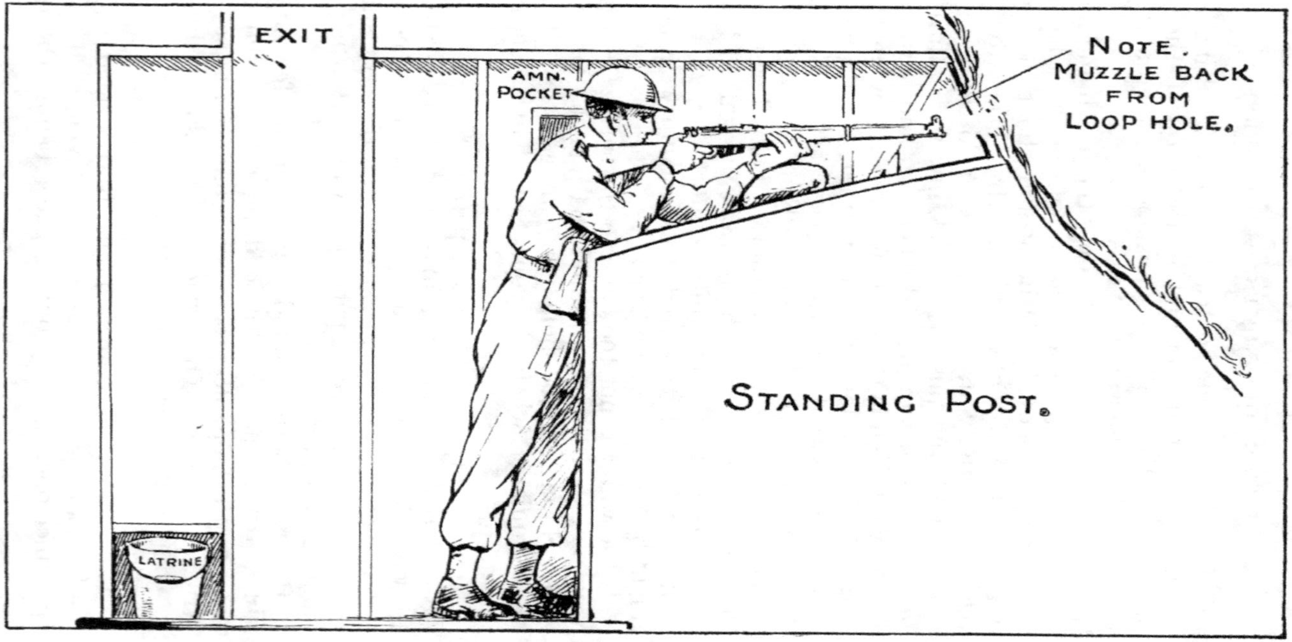
EXIT
AMN. POCKET
NOTE. MUZZLE BACK FROM LOOP HOLE.
LATRINE
STANDING POST.

CONSTRUCTION OF SNIPERS' AND OBSERVERS' POSTS

one, should slide in grooves so that it will move with the rifle at different angles. Shooting straight out of one plate would not be much protection, but at any angle it would be next to impossible for the enemy to put a bullet through both loopholes.

Discarded loopholes should be left, and the sentry go around and darken them up, so that the enemy will shoot at them. In all the trenches the parados should be higher than the parapet; this does away with the silhouette when anyone looks over, and it prevents the blowing back of shells. When much tunnelling and heavy construction is required in connection with this work, the O.C. Snipers should make requisition to the R.E. for the necessary assistance and material.

If a well-constructed Sniping Post is being abandoned for some reason or other, it should be always blown up or otherwise destroyed, so that the enemy may not find out anything about our methods.

A Sniping Post must be occupied all day in cases where you cannot enter or leave it without being seen. It must be occupied before daybreak and vacated after dark, when the listening patrol or guard, in the case of advanced posts, take over. At dusk and dawn, when it is raining, and when a fog suddenly clears away, are most likely times for you to get a good shot at someone.

RANGE CARDS.

The Sniper must know the exact range to any part of his front. His sight will be set for the average range and unless he has time to make sight adjustments he aims up or down for range differences.

The object of your setting ray must be the most prominent point in the enemy's parapet; then you can pick out other prominent objects all along the enemy's parapet, not over ten in your sector, for if you have more than that it will be hard to remember. Now you number them off from left to right, including the setting ray point.

In occupying a post you make a range card for each

loophole if there is more than one. You draw rays on your range card to every point along the enemy's parapet. Do not draw rays to any point that you might be able to see in the rear of the enemy's trenches. By using these definite points and the clock method you can call one another's attention to any given point in your sector. An object directly to the right of point 4 will be indicated as "four three." That is, point 4, 3 o'clock. Use the nearest point to the object as first indicated.

The lower half of the clock face will be rarely used, and the upper half only when you see the ground or objects in the rear of enemy's trenches. Then you call out 10, $11\frac{3}{4}$, 12, 2 or 4 o'clock, as the occasion demands, but 3 and 9 will be the figures mostly used. The method is a sure one and any good marksman knows it by heart for he has constantly used the clock face to indicate the position of shots on the target.

Range cards should be of cardboard or very thin wood, and painted the colour of the surroundings. It should be small enough to be able to place it under loopholes. Never use white paper for range cards.

Officer Commanding Snipers must have a range card for each Sniping Post for reference when he reads the daily reports. In making out a card mark the range on each ray. Any point in the rear of the enemy's trenches you wish to call attention to you mark on the top of the range card, also a description of the particular point you wish to call attention to. Additional information may be written on the back of the card. You can write on both sides of this if necessary. Range cards must always remain in the post to which they belong. They should be explained carefully to Observers and Snipers of a relieving battalion. Precautions must be taken that they do not fall into the hands of the enemy.

Armour Plate and Sandbags.

Armour plates for protection are an important part of the Scout-Sniper's equipment. They are usually

CONSTRUCTION OF SNIPERS' AND OBSERVERS' POSTS

indented for through Brigade and believed now to be made to standard size. The ideal shape would be like a giant "tin hat," so that no matter where a bullet hit on it a curved surface would be presented to the impact. The next best shape is a plate bent to a right angle, but such a plate is much heavier than a flat plate to fill the necessary space.

Steel plates should always be covered with sacking or other material to disguise their nature. They must be placed in position carefully so that only the sniping or observation hole is left uncovered from the front. The hole must be closed when not in use.

When a Sniper's post is very near the enemy the following method can be used. Build an emplacement not less than three sandbags thick, using clay or heavy earth to fill the bags, in front of the emplacement through which you wish to shoot. Beat the bags down thoroughly and let them settle for twenty-four hours. Then dig from the inside through the sandbags, and through the centre of the outer clay-filled sandbag about 6×6 or 4×4 inches, so that you only have the single sacking left on the outside. You can see through the single sacking and cannot be discovered from the front even at 10 yards, as the sack will keep its shape, having been filled and settled before. If you can see through the sacking you can aim and fire through it. For a very short range this method cannot be beaten.

Keep the muzzle of your rifle 4 feet back, so that the discharge will not burst the single sacking. When fired through several times replace that sack with another one, but be careful that you get one of the same shade. This must be done at night. It may be better to build a new post a few feet to one side, and you will have no trouble with different coloured sacking.

If, when you are constructing a Sniping Post, your work is discovered, abandon it, but set a contact mine so that the enemy will be sorry for his discovery. Of course, this only applies to advanced posts. If you can locate an enemy advanced Sniping Post, creep up in the night and capture guard quietly. Take possession. Gather up any of the enemy patrols that night who visit the post. Then mine and abandon before daylight (contact mine). It will rarely be practicable to occupy an enemy Sniping Post. If such a thing were possible it would be only good for one day, as the enemy would in all cases find it out before the second night. If plans are carefully laid and carried out the enemy may lose a good many men over one discovered advanced post. Take precautions that they don't play a similar trick on you.

eighteen

THE CONSTRUCTION OF LOOPHOLES

THERE are so many tricks and devices in the construction of loopholes that pictures are the only way of showing them. This section, then, is mostly pictorial.

Used principally in Even, Well-built Parapets.

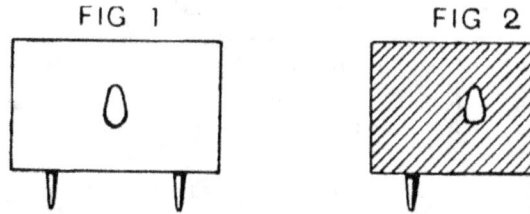

Take an ordinary steel plate. (*Fig.* 1.) Cover plate tightly with sacking or cloth, tightly sewn on. When sewn, cut hole for aperture, so that the shutters will work freely. (*Fig.* 2.)

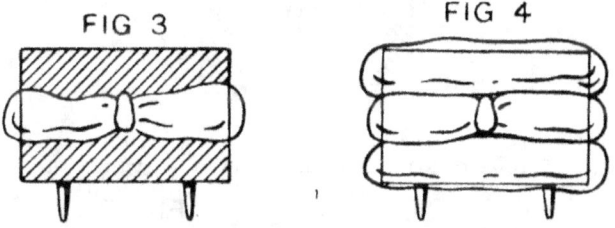

Take two sandbags, and fill them about half-full of rags or old torn sandbags, being particular to note that the corners are filled tightly. Having filled these sandbags half full, turn down the choke ends and sew them up securely. Then sew one on each side of aperture as in Fig. 3, to correspond as near as possible with headers in solid parapet. (*Fig.* 3.)

Take two more sandbags for stretchers, and fill them as tightly as possible with old sacks or rags ; fill them to correspond as near as possible with the stretchers in permanent parapet. Then sew these stretchers securely to plate. (*Fig.* 4.)

SNIPING, SCOUTING AND PATROLLING

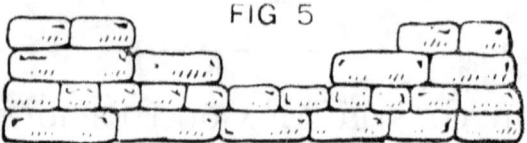

Fig. 5.—Parapet prepared to receive Plate.

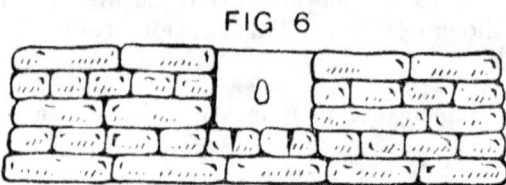

Fig. 6.—Plate in position.

To prepare this plate successfully requires a certain amount of practice, particularly in filling the bags and sewing them up. (*Fig. 6.*)

No. 2 Concealed Plate ("Fitzgerald").

(*Can be used in uneven or broken parapet.*)

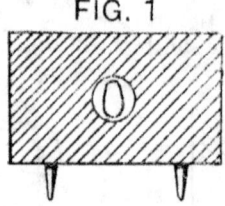

Fig. 1.—Steel Plate with sacking sewn tightly over it and a Hole cut to enable Shutter to work freely.

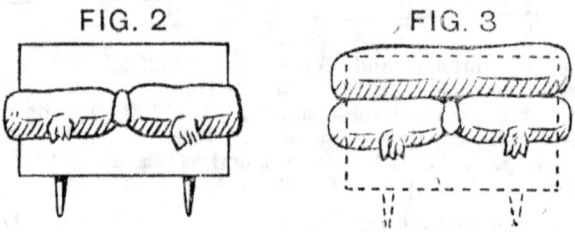

Fill up choke end of two bags about 10 inches in depth, or more if necessary. Fill fairly tightly, turn down, and sew up the ends, and sew these two dummy chokes to the plate. It is sometimes necessary to sew on three chokes in order to cover the plate completely. (*Fig. 2.*)

THE CONSTRUCTION OF LOOPHOLES

Fill and sew false stretcher to plate, being always careful to see that all parts of the plate are covered from front view. (*Fig. 3.*)

NOTE.—To obtain the best concealment with this plate, the dummy stretcher on the bottom of the plate should be dispensed with, and the plates placed low enough in the parapet to allow the false chokes to remain on the ground-level, as in Fig. 4.

FIG. 4.—SHOWS PLATE IN POSITION IN PARAPET.

The steel lugs are driven into a filled sandbag for support.

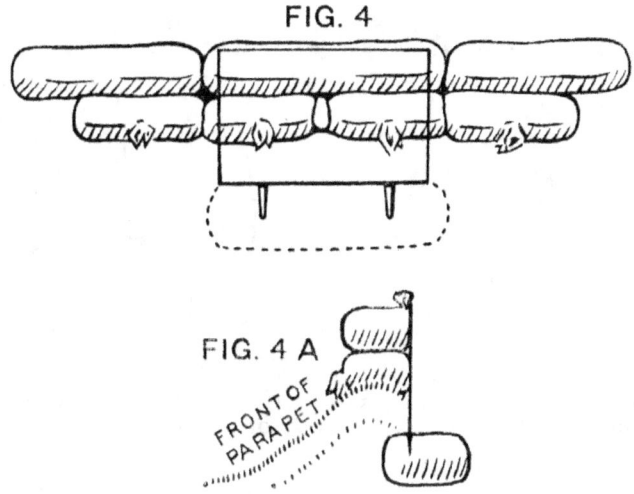

For better protection, and to gain more room for the body, this plate should have two steel wing plates; this requires more work, but is much better than sandbag protection. (*Fig. 5.*)

FIG. 5.—WING PLATES IN POSITION.

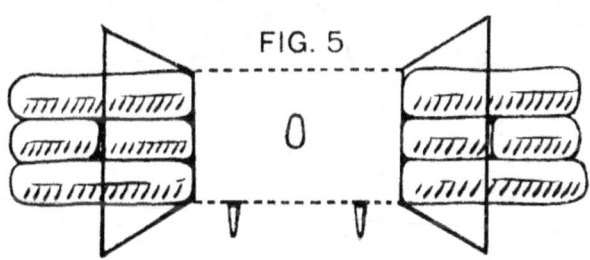

LOOPHOLE IN HEADER.

Place plate flat on the ground with front of plate uppermost. Then obtain piece of tin 10 inches and roll it to a diameter of 6 inches, and fasten so that it will spring open. Place this tin over the aperture and then get header sandbags and cut off the required length, and place over tin. Sew to sacking or plate. Then sew to shaded part (*Fig.* 1) between tin and bag, and pass stiff piece of wire through or around (threaded) neck of bag to keep it open. (*Fig.* 1.)

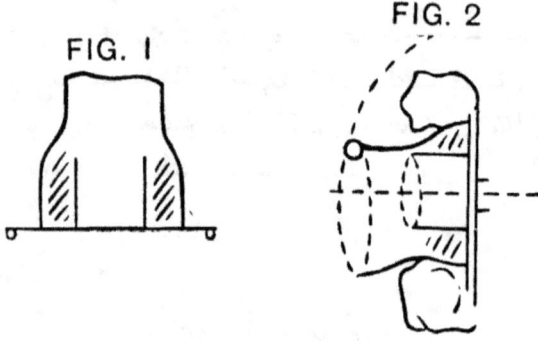

Fig. 2.—Cross Section, showing small Weight fastened under Upper Lip of Sack, and Loophole open in readiness to Fire.

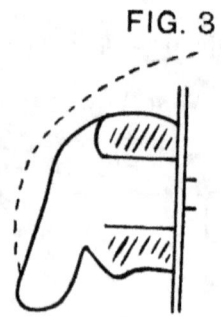

Fig. 3.—Loophole closed.

THE CONSTRUCTION OF LOOPHOLES

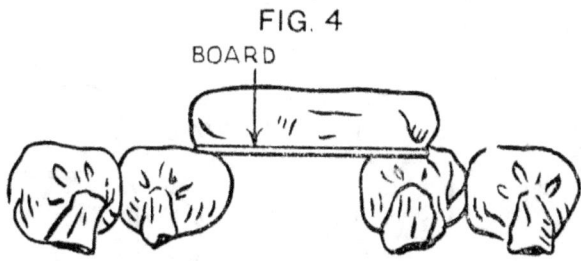

Fig. 4.—To prepare Parapet.

This loophole can be best concealed by being placed on ground level; therefore leave space for false header, and place board over aperture, resting on headers as in Fig. 4. To support real stretcher, it is not necessary to have false stretcher. Plate in position with loophole open; wire passed over top of parapet. (*Fig. 5.*)

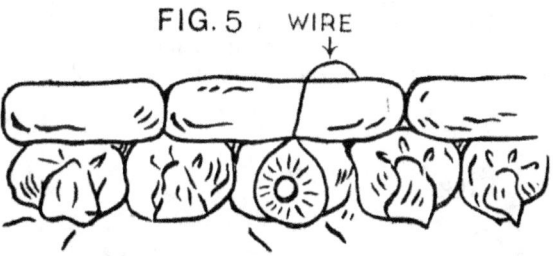

LOOPHOLE BETWEEN TWO ORDINARY HEADERS.

This loophole can be used in any parapet where bags are placed with the choke end outwards. This loophole is very difficult to pick up if properly manipulated. Steel plates can be used if required.

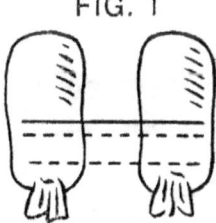

Place two header bags about 6 inches apart in the parapet. Then place a piece of board 1 inch by 6 inches across these bags, as in Fig. 1. Then take an empty sandbag and fold it to an oblong shape, about 6 inches wide, and sufficiently long to cover aperture effectively, as per dotted line in Fig. 1. You are now looking down on top of bags. (*Fig. 1.*)

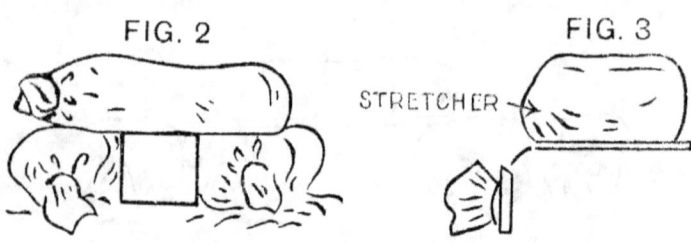

Now place genuine stretcher bag across aperture, and supported by 1 inch by 6 inch board, and also holding in place the dummy sacking, as in Fig. 2. (*Fig. 2.*)

Behind the sacking shutter is nailed a piece of wood about $1\frac{1}{2}$ inches thick to keep shutter steady, and in front is nailed to the wood a false choke. To this false choke is attached a piece of signal wire and passed over the parapet. (*Fig. 3.*)

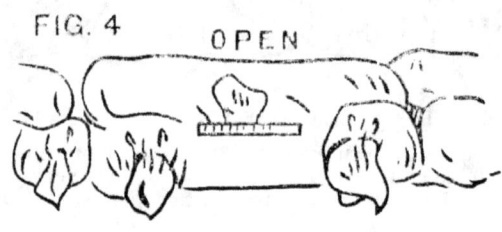

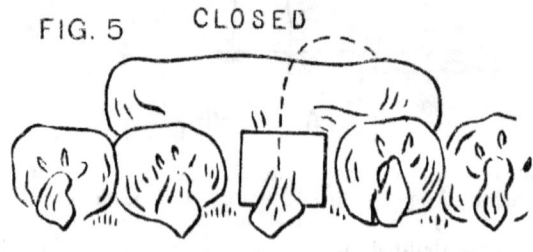

THE CONSTRUCTION OF LOOPHOLES

MOUSE-TRAP LOOPHOLE.

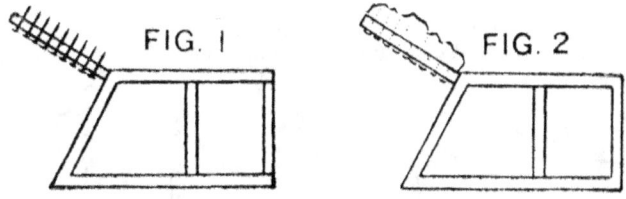

Mouse trap with lid open, showing the nails driven outwards through the wood. On these nails can be placed sod, either with sod downwards or upwards, or with anything placed on to agree with the outward appearance of the parapet. This frame can be used, or an ordinary box can be cut to the correct angle. (*Fig. 1*.)

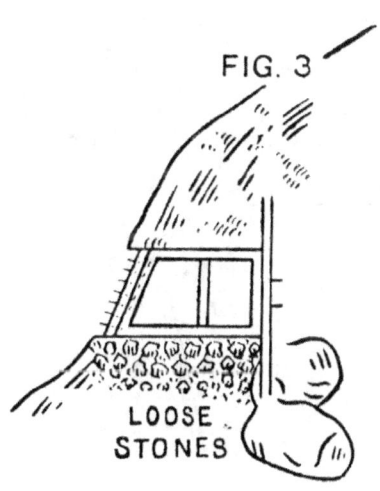

Shows the steel plate in position with the mouse trap pressed close to the loophole in the plate, in readiness to receive the required disguise on the nail-points. (*Fig. 3*.)

SNIPING, SCOUTING AND PATROLLING

LOOPHOLES IN PARADOS.

Fig. 1.—Small Wooden Box placed in Stretcher Sandbag in Parados.

Difficult to observe, and to be used as a reserve loophole, to be fired from occasionally. Plate can be used here if necessary.

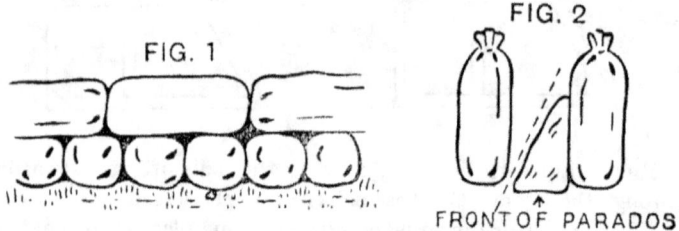

Fig. 2.—Looking down upon Parados.

Sew up a bag into three-cornered shape, as per Fig. 2, and place it slightly forward from flush to leave aperture for enfilade fire. Plate can be used if necessary. This loophole may be used with success in front-line parados. (*Fig.* 2.)

Observation Box.—This box can be placed on top of parapet, let in about 1 foot or 18 inches, and gives good view, not easy to pick up; several of them can be used on parapet. Small pieces of tin painted can be used as a shutter. Forward aperture, about 4 inches by 6 inches; rear aperture, about 14 inches. This box can also be used for observation in the parapet if made smaller.

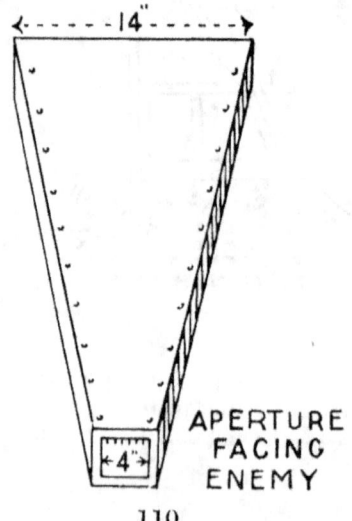

THE CONSTRUCTION OF LOOPHOLES

LOOPHOLE FOR OBSERVATION.

Take a new sandbag and cut along dotted line—one side only, leaving other intact. (*Fig. 1.*)

Turned around, the bag now looks like a shaving-case, with pocket at bottom or header end. This pocket is now filled with old rags or bags to represent a dummy header. When carefully filled, a piece of board of the same shape as the header is placed in behind the rags sufficiently deep to allow the edge of the pocket being pulled across and sewn to remaining side of sandbag. One side of the sandbag is left intact in order to hold the dummy header in position, and also to act as a hinge for opening and shutting. (*Fig. 2.*)

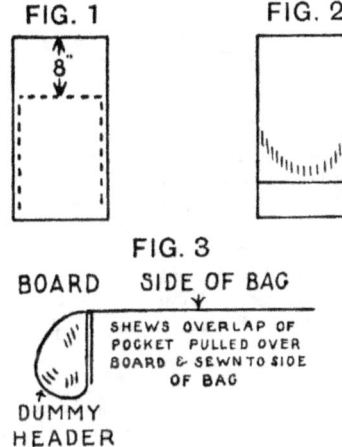

Fig. 3.—SIDE VIEW OF DUMMY HEADER, AND REMAINING SIDE OF SANDBAG IN READINESS TO BE PLACED IN POSITION.

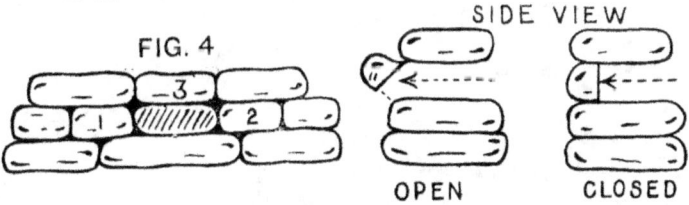

Fig. 4.—DUMMY IN POSITION IN PARADOS, REPRESENTED BY SHADED BAG.

A piece of board is placed across sandbags 1 and 2, in order to support stretcher bag No. 3, and at the same time to hold dummy in position. (*Fig. 4.*)

BOX LOOPHOLE FOR EARTH PARAPET.

Fig. 1.—Ground Plan of Box, showing position of Plate and Field of Fire.

Box can be short or long, according to circumstances; the shorter the better. The front of the box should be cut at an angle, in order to conform to the slope of the parapet. (*Fig.* 1.)

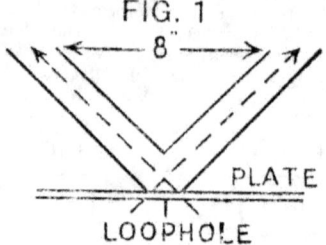

The dimensions of the box depend largely upon conditions of the parapet and opportunity for concealment. If careful concealment is necessary, the box aperture should not be larger than 4 inches by 3 inches. These apertures should be closed by means of a wooden door of the same size. This door can be hinged either from side or top, and opened by wire attachment. Concealment can be arranged in usual manner by driving nails through door, and placing on these nails anything which conforms to the background or general colour of parapet. (*Fig.* 2.)

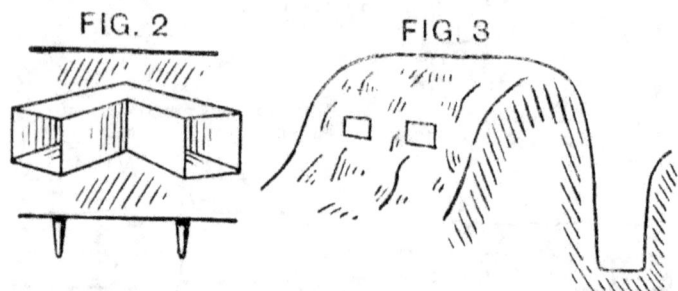

Fig. 2.—Front View of Box and Plate in readiness for Parapet.

Fig. 3.—Section of Earth Parapet showing the Loophole in position, but not yet disguised.

Small sods grass downwards, and placed on the abovementioned nails, provide very good concealment. This loophole gives a very much larger field of fire than the ordinary aperture, and is safer.

nineteen

DECOYS AND THEIR USE

WE wish to impress on you that there is nothing in which you are so liable to go too far or overdo as in constructing, and especially in manipulating, decoys. Don't overdo it, whatever you do. You can always go a little farther along the same line if you find it necessary, but if you overdo it all your work is lost, not only for you but perhaps all along the line. Decoys need common sense in both construction and use.

Get hold of some old clothing and rifles that are beyond repair. You can easily make a figure to represent a man by using a uniform stuffed with hay or straw. Put heavy wire or wood inside to make it rigid. Head and face can be made out of factory cotton, stuffed with hay or straw. You make nose and the rest of the body by making the material to suit. Paint the eyes, make ears, and sew them on. Hair and beard, if necessary, can be made of horse-hair or cow-hair of suitable colour and stuck on with paint or glue.

Hair may also be made with unspun flax. Most perfect heads can be cast in plaster, head and all painted to suit, but they are very fragile. The closer to the enemy's lines you are, the more perfect-made decoys are required. You must make decoys to represent English, French and Imperial soldiers. By exposing a French soldier, when they think they know the English are occupying a position, confuses the enemy and makes him eager to find out things, and perhaps expose himself in doing so. It is up to you to make decoys look interesting to the enemy.

When you are operating close to enemy lines you must have your decoys well made and manipulated, as he has powerful telescopes. You can often place a figure in a tree, having one string or wire on each side to move the figure slightly. Again the prospects are that the enemy will expose himself. You can also place figures in the grass or undergrowth in No Man's Land, and move them slightly by strings to attract his attention. When you expose figures over the parapet, you show only the head and, in some cases, the shoulders; and be sure to have figure duck every time it is shot at, whether it is hit or not. A figure of the head of a man with dummy field-glasses, looking over a parapet, will nearly always draw fire. Decoys are rarely of any use unless they are movable or operated by strings, sticks or wires from some safe place.

In operating a line of decoys it must be borne in mind that artillery, as well as machine-gun and rifle fire, may be used by the enemy, so operators must govern themselves accordingly. Don't be too close to your decoys unless you are well dug in.

Another worth-while trick is to creep out in the morning before daylight and put a few tin cans close in to his wire or trenches. Have a strong wire or string attached hid in the grass. By rattling the cans you can, perhaps, excite his curiosity and get a good shot at him, particularly if it is raining heavily and the trenches are a good distance apart, and your Snipers are occupying an advanced post. You can also use a looking-glass to flash light into his trench, and in this way arouse his curiosity, and when he looks over to see what is going on again you have got your chance. Always look out that the enemy does not play a trick like this on you. Anything that you want to look at do so from a concealed place. You can also have hand-grenades placed close to the enemy's lines and arrange to explode them by pulling a string. If your trenches are located beyond hand-grenade range, it will most certainly excite his curiosity, particularly at dusk or dawn, heavy rain or fog. In practising the

DECOYS AND THEIR USE

above ruses where there are likely to be several Huns showing themselves at one time, it will be well for the observer to be ready to shoot. Also in case of an attack they should both shoot, but in all other cases only one man shoots from a post. You can also in many cases make fake Sniping Posts outside the lines and mine them, and when enemy patrols come to investigate blow them up. Such fake posts should be open-top ones.

By arranging a solid head decoy it is possible to locate an enemy sniper by the direction of the shot-hole through it. Or a screen can be arranged behind the decoy head so that the direction from which the shot came can be discovered by aligning the hole in the head and the hole on the screen. Naturally, the utmost care must be taken not to expose your own head in your endeavour to align the holes on the Sniper. A periscope might be arranged to be put up exactly where the hole had gone through the head. Of course a decoy head is pulled down directly it is hit in most circumstances, but in this case the head may remain up. The enemy Sniper may think he has missed and try another shot whilst your periscope is directed at him.

twenty

SQUIRT GUNS

LIGHT automatic rifles or large auto-pistols are being used extensively in most armies as infantry weapons in situations where portability and a heavy volume of fire at short range on the instant are necessary.

Desirable features in such weapons are shortness and lightness, so that they will not impede men on patrol who may have to crawl several hundreds of yards. Their rate of fire should be very high and the ammunition compact and light.

The best known of these weapons is the Thompson sub-machine gun. They became notorious during the gangster period in the United States and from that fact the whole class is known among English-speaking peoples as " gangster guns " or by the American slang names of " squirt guns " or " Tommy guns." The Thompson fires an automatic pistol cartridge, weighs about 11 lb., which is heavy for war work, and has a rate of fire of about 1,500 rounds a minute. Though listed as a United States army weapon, it has been displaced by the Garand semi-automatic rifle.

The first step in the development of the squirt gun was taken by the Germans in 1917, when they fitted a drum magazine carrying 32 rounds to the long Luger-Parabellum pistol. At the end of the Great War they were using a machine pistol or automatic carbine, the Bergman Muskete. Their present " patrol pistol " is of much the same design. It was this light arm, or a weapon of similar design, which was used recently by the Finns at close quarters against the Russians.

During the last war the Italians developed the Villar-Porosa, an ingenious and interesting double-barrelled machine gun-pistol in which the barrels could fire singly or both together. With both barrels in action the rate of fire was about 2,400 rounds a minute. The adoption of this weapon by the British Army was considered, but possibly its weight, over 13 lb., was against it.

Firing from the Hip.

The multiplying of infantry weapons is not desirable, but a rifle has not a sufficiently high rate of fire to deal effectively with situations which may arise in patrol work or raids.

A light automatic such as the Bren can be used, and is used, in patrolling, but it is heavy where much crawling has to be done, and may be too slow in coming into action. A " squirt gun " weighing 10 lb. or less is not difficult to crawl with and can be fired from the hip or shoulder. If of good design it can be brought into action almost as quickly as a pistol ready in the fist.

The United States Ordnance Department believes that it has solved the problem of providing an all-round infantry weapon in the Garand semi-automatic rifle. This ingenious arm weighs 9 lb., can be fired single shot or at a rate of over 100 rounds a minute, including recharging the magazine after every eight shots. Its length is 42 in., that is, $1\frac{1}{2}$ in. less than our S.M.L.E. It is, however, about 5 oz. heavier than our Service rifle. Certainly, the Garand is a good compromise between magazine rifle and light machine-gun, but it is doubtful whether, with a magazine holding eight rounds, it would be useful in a sudden bump against the enemy that might happen to a night patrol. The man who has thirty to fifty rounds in his magazine can make a couple of sweeps of fire in the direction he thinks the enemy is, and will probably cause casualties to him. Eight rounds are gone almost before the rifle can be moved.

It would seem that the squirt gun is a necessity, and that it will be increasingly used in the armies of the belligerents.

twenty-one

THE TELESCOPIC SIGHT

A TELESCOPIC sight will not make a poor shot into a marksman nor will it turn a bad rifle into a " gilt-edged " weapon. The virtue of the telescopic sight is that it enables an expert shot to take the fullest advantage of his abilities, to aim accurately at objects he can scarcely see with the naked eye and to shoot with effect in bad lights. A good rifle and a good shot can give of their very best when they have the aid of a telescopic sight.

An ordinary telescope consists of two or more systems of lenses so arranged that the light reflected from distant objects is gathered in and the resultant bright image magnified so that the distant objects appear to be nearer the eye than they actually are.

The magnifying power of a telescope is usually indicated by a figure with the multiplication sign after it. If the figures and sign are " $20 \times$ " it indicates that the particular instrument has a magnification of 20 and means that the object is apparently made twenty times bigger.

The telescopic sight is an ordinary telescope with the addition of a pointer which acts as an indicator for aiming the rifle to which the telescope is rigidly fixed.

Many men used to the ordinary open battle sights become confused in an endeavour to find out which is the backsight and which is the foresight. A simple way of overcoming this difficulty is to put an ordinary telescope on to a fixed rest so that some definite and well-defined object, a big building, a church or a windmill, is exactly in the centre of the picture. Explain that the telescope is aimed at that object and that all the pointer in the telescopic sight does is to indicate the exact point of aim. There is no such thing as a backsight or a foresight in the telescopic sight as the whole lens system acts as a sight.

As has been said it is necessary that the telescopic sight should be rigidly fixed to the rifle. No explanation of this is necessary beyond the fact that ordinary battle sights are rigidly fixed. The student will be already familiar with the working and use of ordinary sights and should appreciate the necessity for a rigid mounting. The pointer or indicator is placed in the optical focus of the front lens (objective). It is then perfectly sharp. Also, by this arrangement, no matter what the relation of the eye to the eyepiece the pointer remains always in exact relation to the mark on which it is aimed. To demonstrate this put the rifle in an aiming stand with the pointer aligned correctly at six o'clock on the aiming mark of a target. Ask the student to look through the sight and move his head about so that the relation of the eye to the eyepiece is changed. He will see for himself that no matter where his eye is the pointer is always directed on the mark.

If the pointer appears to wander as the eye is moved then the fault known as parallax is present in the instrument and it should be returned to Ordnance for adjustment.

As the telescopic sight is rigidly attached to the rifle means for altering the elevation are necessary. Elevation is changed by moving the pointer up or down as necessary. On the top of the Service sight there is a knurled drum about as big as a halfpenny on the upper surface of which the range is marked out in divisions indicating hundreds of yards. By twisting this the sight can be set to any desired elevation. A locking screw is provided and should always be used. Screw it up tightly after each sight adjustment.

Means of adjusting the lateral deviation of the sight is also provided. This is in the form of a small wedged-shaped prism at the object glass end of the telescope.

THE TELESCOPIC SIGHT

By removing the front cover tube a graduated brass ring is exposed. A very small movement of this ring is usually all that is necessary to bring the point of aim coincident with the axis of the barrel. If the combination of rifle and sight is shooting to the left the graduated ring is moved to the left or counter-clockwise. If the shots are striking to the right the ring is moved to the right or clockwise. The adjusting ring is marked " 5," " 10 " etc. Each division represents a movement of 5 inches at 100 yards.

On the underpart of the body of the sight there is a small screw head which can be freed to adjust the focus of the telescope to the eye of the user. The moving part is adjusted to exact focus and the screw is then tightened.

These three adjustments are the only ones allowed to be made by the Scout-Sniper or the Armourer of his Battalion. Stripping down the sight is absolutely forbidden, and if any adjustments other than those indicated are required the instrument must be returned to Ordnance for the attention of a properly qualified optical instrument maker.

The telescopic sight is a delightfully easy and simple thing to use. There is absolutely no strain on the eye and since the magnification of the Service sight is low, usually $3\times$, the wobble or tremor of the marksman is not much exaggerated. To use the sight all that must be done is to look through it—any old how—and place the pointer just below the object to be hit. Aiming off for wind movement is very easy with this sight.

The sight is so arranged that it has a large field of view—about 45 feet at 100 yards—and it is also so designed that the full field is available when the eye is anything within about two inches from the eyepiece. This " relief," as it is called, is necessary to prevent the eye being damaged by the recoil of the rifle and sight. Some men fix a piece of rubber tube over the eyepiece to cut out the light between the lens and the eye. This may be done and should be tried. Similarly a piece of rubber tube may be attached to the object glass end of the sight to prevent the sun reflecting from the lens. This expedient is only necessary when the sight is being used in the open. Normally, of course, the Sniper will be in a hide when the glass will be well protected from a front sun.

The Service telescopic sight is stoutly made and will stand considerable rough usage. It should, however, be treated with respect and care, and always removed from the rifle when not in actual use. It should be carried in the stout leather case provided which should be slung over the shoulder.

If the lenses become dusty or dirty or spotted with rain they should be carefully cleaned with a small piece of chamois leather or, if this is not available, with a piece of clean rag or a handkerchief.

" 4 by 2 " should not be used for cleaning the lenses, as it is often gritty and usually slightly oily.

twenty-two

CONCLUSION

IN conclusion, we wish to say that these notes do not by any means cover all the Scout-Sniper should know, nor his training and organization. Everything written is based on experience gained during the war of 1914-1918. The notes have been brought up to date where possible, but no training notes of new methods are yet available. Probably they are not yet necessary for nothing much that is new in Scouting or Sniping has yet developed. New things will come, of course.

This book should serve as a foundation to start from, and the intelligent Scout-Sniper on active service will be able to add a great many useful ideas, as he is confronted with the various conditions in the discharge of his duties.

The more one goes in for this work the more interesting it is, and the more one realizes the immense importance of this branch of the Service. It is an extremely deep game, and should be played to the utmost limit. If entered into on these lines, a great many lives and much ammunition and other war material will be saved by the end of the war that would otherwise be lost.

We have lots of big-game hunters and marksmen in our Army to draw from for this work, who, with a little special training along these lines and a proper organization, can make enemy Sniping impossible and life in their front line not worth living. Our Infantry as a whole can easily supply five per cent. of its total who can be trained as effective Snipers and Scouts.

We think that every marksman should be trained in Sniping and Scouting, so as to have lots of men to draw from for this important work. After completion of their training they should be graded and employed according to their individual ability.

If a unit has enough shots for its full complement of Snipers, let the others work with their companies as usual. They will make excellent men to have along the firing line at any time, and will replace Sniper casualties as required. We cannot have too many of such trained men, even if they are not actually employed as Scout-Snipers for the time being. The enemy have a regular system and organization regarding this work. They have the Schutzen battalions, the members of which are drawn from all over their country. The men in the reserve are all members of Rifle Clubs and have kept up their marksmanship for years. It is up to us to have at least the same or better, as we have far better men to draw from. As a Scout-Sniper you are thrown on your own resources to a large extent, and work away from your superiors' immediate supervision. You must be a man in possession of a very strong sense of honour, and must feel the responsibility and trust placed in you, and not neglect your duty at any time because you may think you have an opportunity to do so. Yours is an all-the-time job; a few moments' neglect of your duty may be fatal to yourself and others. The more trust your superiors place in you the more responsibility you accept. If you don't do your utmost, you are helping the enemy to the extent of your neglect.

Even in comic papers of to-day you notice new and original jokes. Stale jokes are of no use. So it is in this specialist branch. Your mind must be continually concentrated to invent something new, as stale and old-fashioned methods are useless, and perhaps fatal, in modern warfare.

This war is not going to be won by ancient rules and regulations, but original and up-to-date methods, and the methods employed in this branch lie entirely with the adaptability and capability of the individual Scout-Sniper.

PLATE XIX.—THE SNIPER WITH HIS P. '14 RIFLE AND TELESCOPIC SIGHT. He is hidden from the front. Compare with Plate XX.

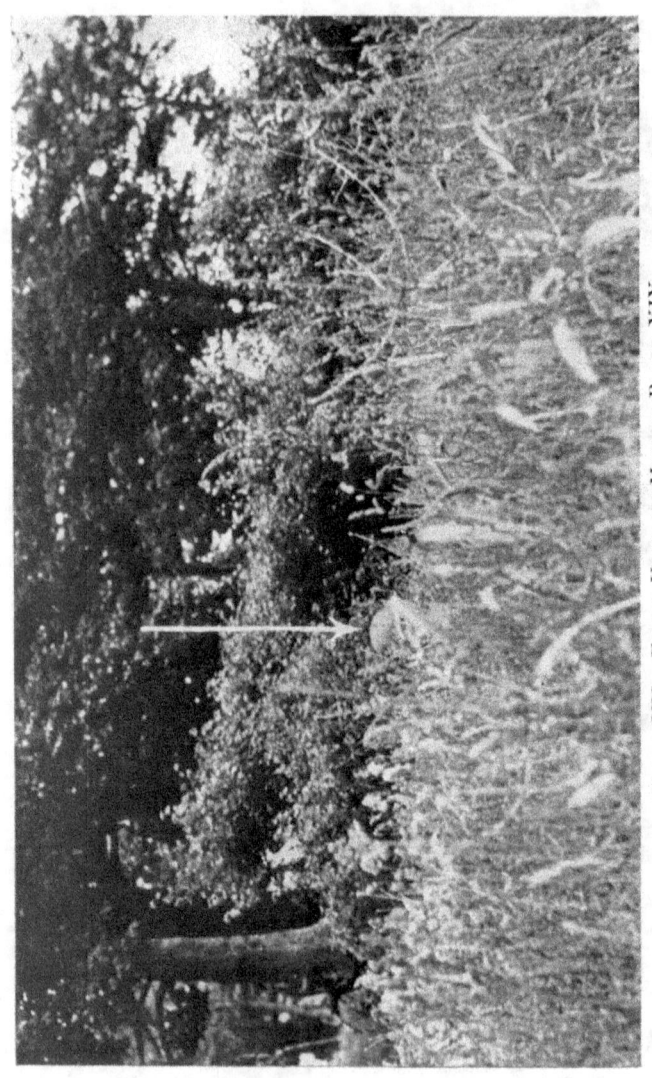

PLATE XX.—FRONT VIEW OF MAN IN PLATE XIX.
With stockinette cap or grass over steel helmet he would be quite invisible.

PLATE XXI.—CAMOUFLAGE SCREEN.

PLATE XXII.—CAMOUFLAGE SCREEN SEEN FROM FRONT.
Compare with man lying in open.

PLATE XXIII.—CAMOUFLAGE MADE OF NET AND RAGS.

PLATE XXIV.—CAMOUFLAGE OF NET AND BRACKEN.
(See Plates XXV and XXVI.)

PLATE XXV.—MEN IN CAMOUFLAGE NETS.
Compare with centre man.

PLATE XXVI.—SAME THREE MEN SEEN FROM DISTANCE.

PLATE XXVII.—Man on Right shows use of Camouflaged Head.

PLATE XXVIII.—SNIPER MAKING USE OF NATURAL COVER.
Only steel helmet gives him away from distance of 10 feet.

PLATE XXIX.—TYPE OF DISAPPEARING TARGET USEFUL FOR SNIPER PRACTICE.
Represents machine gunners. Worked from forward trench.

www.ingramcontent.com/pod-product-compliance
Lightning Source LLC
Chambersburg PA
CBHW050206130526
44591CB00035B/2269